Living Things

V. Slaughter

Hodder & Stoughton
A MEMBER OF THE HODDER HEADLINE GROUP

Preface

Living Things covers the material common to GCSE biology syllabuses and the biology component of GCSE science syllabuses. By presenting information, concepts and principles visually I hope this book will help all pupils to understand the subject and prepare for an examination.

Each chapter has the same structure so that the reader will quickly gain the confidence needed to use the book effectively. A chapter consists of short passages which are easy to read, simple clear diagrams to explain and convey information, comprehension exercises, a summary, a list of key words accompanied by definitions and questions which demand recall, reasoning and interpretation.

By using concise text this book should encourage the development of basic skills as well as the acquisition of facts which are likely to be examined. Short comprehension exercises can be used to reinforce the most important points included in each section of a chapter. The summary at the end of a chapter focuses attention on fundamental principles and concepts. A glossary of key words used in a chapter consists of the words and biological terms which are often found to be difficult to understand and use correctly. Questions at the end of a chapter have been designed to develop techniques needed to cope with formal examinations.

Wherever possible, questions based on the practical work, which normally forms an integral part of a biology course, have been included. Although it is hoped that a student will recognize and understand the procedures, results and conclusions through experience in a laboratory the questions contain sufficient information and data so that they could be answered by reading and understanding the chapter.

V. Slaughter

Orders: please contact Bookpoint Ltd, 39 Milton Park, Abingdon, Oxon OX14 4TD. Telephone: (44) 01235 400414, Fax: (44) 01235 400454. Lines are open from 9.00 - 6.00, Monday to Saturday, with a 24 hour message answering service. Email address: orders@bookpoint.co.uk

British Library Cataloguing in Publication Data
Slaughter, V.
  Living Things.
  I. Biology
  1. Title
  574   QH308.7

ISBN 0 713 10416 3

First published 1980
Impression number 29 28 27 26 25 24 23 22 21 20
Year   2004 2003 2002 2001 2000 1999 1998

Designed by SGS Education, 8 New Row, London WC2N 4LH. Illustrations by Barrie Thorpe. Printed in Great Britain for Hodder & Stoughton Educational, a division of Hodder Headline Plc, 338 Euston Road, London NW1 3BH by Scotprint Ltd, Musselburgh, Scotland.

Contents

Acknowledgements

The Publishers would like to thank the following for permission to reproduce copyright photographs:

Heather Angel: 191

S. Beaufoy: 188, 191

John B. Free: 26

Institute of Geological Sciences: 186

London Scientific Films: 66(b), 66(c)

A Shell Photograph: 66(a)

WHO Photograph: 421, 42r

1: Living things

Biology is the study of living things. The microscopic plants and animals found in pond water are very small living things. The giant red wood tree and the blue whale are very large living things. Although these living things look different from each other they have some **characteristics** which make them alike. They all have the **characteristics of living things**.

Feeding

All living things feed. They need food for **energy**. Plants and animals feed in different ways.

Green plants make their own food. They need sunlight to make sugar from carbon dioxide and water.

Animals eat plants or other animals.

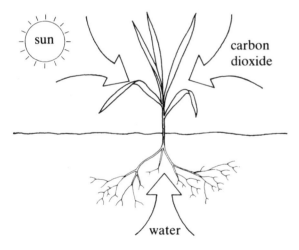

Respiration

All living things respire. When living things respire **energy is released from food**. The energy is used for moving, growing and repairing the body.

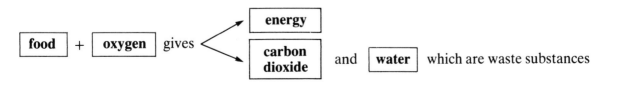

food + oxygen gives → energy

→ carbon dioxide and water which are waste substances

Excretion

All living things excrete. Waste substances must be removed from the body.
Excretion is the removal of waste substances from the body.

A tree stores waste substances in its leaves. The waste substances are removed when the leaves fall off.

Carbon dioxide and water are removed when an animal breathes out.

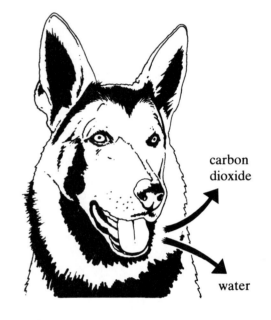

carbon dioxide

water

Growth

All living things grow.

A plant keeps growing until it dies.

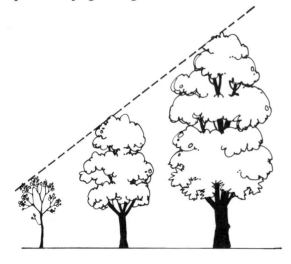

An animal grows until it reaches a certain size. It then stops growing.

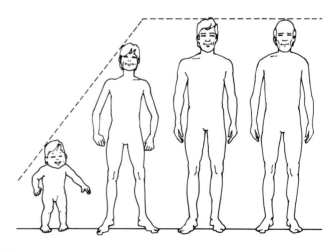

Movement

All living things move.

Some **parts** of a plant can move. Plant movements are very **slow**.

leaves turn towards light

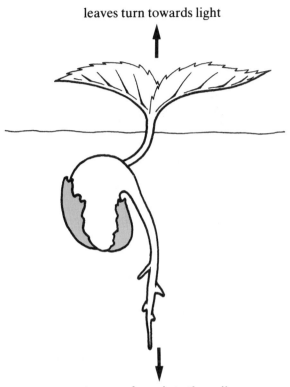

roots grow down into the soil

Animals usually move their **whole bodies**. Most animals can move from one place to another.

a bird uses wings to fly

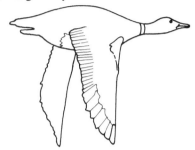

a dog uses legs for running

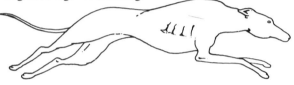

a fish uses fins for swimming

Sensitivity

All living things are sensitive. They **detect** and **respond** to changes in their surroundings.

Plants grow towards light.

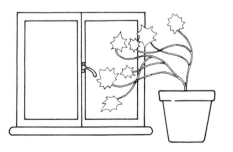

A dog detects his food is ready. He responds by running towards the food.

3

Reproduction

All living things reproduce.

In spring, the horse chestnut tree has flowers.

In spring frogs mate.

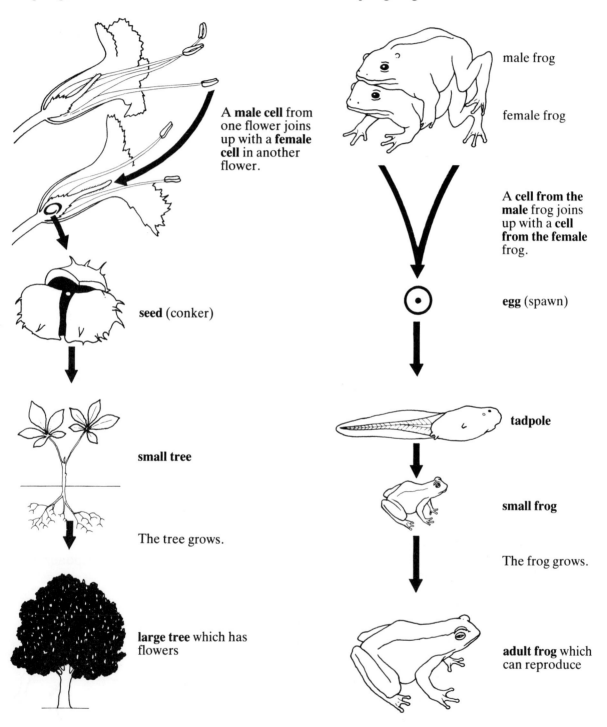

A **male cell** from one flower joins up with a **female cell** in another flower.

male frog

female frog

A **cell from the male** frog joins up with a **cell from the female** frog.

seed (conker)

egg (spawn)

small tree

tadpole

The tree grows.

small frog

The frog grows.

large tree which has flowers

adult frog which can reproduce

Summary

Plants and animals are called **living organisms**. Feeding, respiration, excretion, movement, sensitivity, reproduction and growth are the **characteristics of living organisms**.

Key Words

excretion   The removal of waste substances

organism   A living thing

respiration   Releasing energy from food

sensitivity   Detecting and responding to changes

Questions

1 Look at the plants and animals around you. Choose one plant and one animal to study (remember you are an animal). Copy the table into your book. In each column, write one sentence about the characteristics listed.

characteristic	plant e.g.	animal e.g.	car
1 Feeding 2 3 4 5 6 7			

2 Is a car living? Give reasons for your answer.

3 Name three things which a green plant needs to make sugar.

4 An animal may have fins, wings or legs to help it move. Explain how the following animals move: eagle, mouse, rabbit, bat, goldfish, sparrow, ladybird, shark, penguin, monkey, ant.

5 When an animal respires the waste substances carbon dioxide and water are formed. How are these waste substances removed from the body?

2: Cells

Bricks are non-living things. Bricks are the units which make up walls, houses and other buildings. In 1665 Robert Hooke put some cork under his microscope. The cork was made of 'brick-like' units. Hooke called each 'brick' a **cell**. Cells are the units which make up all living things.

A microscope is used to **magnify** cells. The more you magnify cells, the more you see. The parts of a cell can be seen under the high power of a microscope.

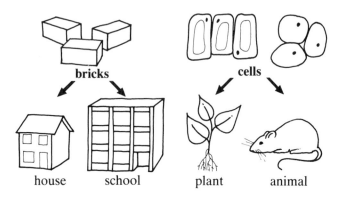

bricks

cells

house   school   plant   animal

Animal cells

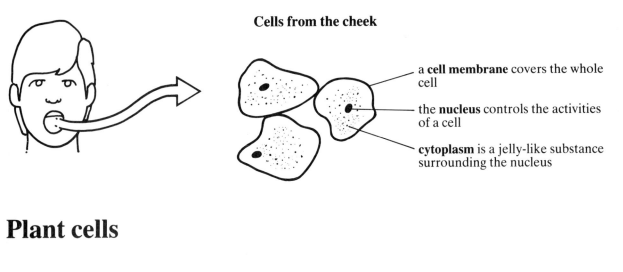

Cells from the cheek

a **cell membrane** covers the whole cell

the **nucleus** controls the activities of a cell

cytoplasm is a jelly-like substance surrounding the nucleus

Plant cells

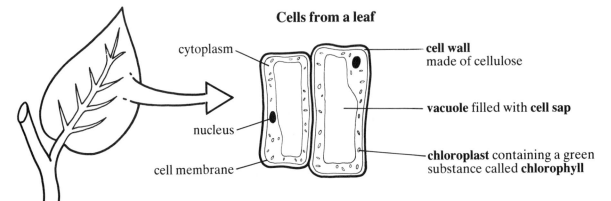

Cells from a leaf

cytoplasm

nucleus

cell membrane

cell wall made of cellulose

vacuole filled with **cell sap**

chloroplast containing a green substance called **chlorophyll**

The differences between plant and animal cells

Although plant and animal cells are similar, there are some differences.

Plant cells	Animal cells
1 have **cell walls** made of cellulose	do not have cell walls
2 have **chloroplasts**	do not have chloroplasts
3 have a **thin lining of cytoplasm**	most of the cell is cytoplasm
4 have a **vacuole** filled with **cell sap**	never have a vacuole filled with cell sap

Questions

1 Are cellulose cell walls found in (**a**) animal cells or (**b**) plant cells?
2 Which part of a plant cell contains chlorophyll?
3 Which cells have a vacuole filled with cell sap?
4 Give two ways in which plant and animal cells are different.

Molecules

Useful substances must go into each cell. Waste substances must come out of each cell. Substances must move from one part of a cell to another.

All substances are made up of many, tiny moving **molecules**. Substances may be either solid, liquid or gas. Ice, water and steam are chemically the same. It is the arrangement of the molecules which makes them look different.

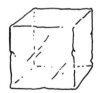

ice is a solid

water is a liquid

steam is a gas

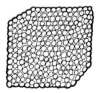

The molecules are tightly packed and do not move.

The molecules can move and there is more space between the molecules.

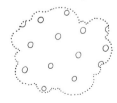

The molecules move a lot and there is a large space between the molecules.

An experiment to show that molecules move

① A purple crystal was put in a dish of water.

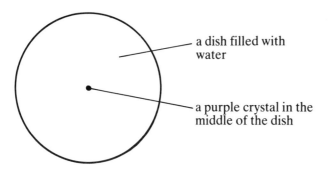

- a dish filled with water

- a purple crystal in the middle of the dish

② At the start of the experiment the purple molecules were **concentrated** in the middle of the dish.

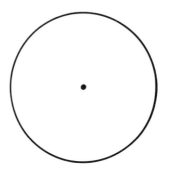

③ **After 10 minutes** some purple molecules had moved.

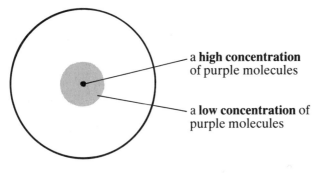

- a **high concentration** of purple molecules

- a **low concentration** of purple molecules

This difference in concentration is known as a **concentration gradient**.

④ **The next day** there was no concentration gradient.

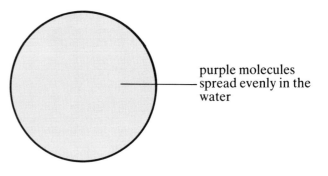

purple molecules spread evenly in the water

The purple molecules spread out by **diffusion**. The molecules moved from a high concentration to a low concentration until they were mixed evenly.

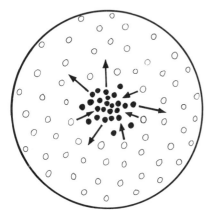

● Purple molecules moved into the spaces between the water molecules.

○ Water molecules moved into the spaces between the purple molecules.

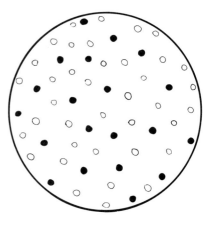

Diffusion

Molecules move from a high concentration to a low concentration until they are mixed evenly.

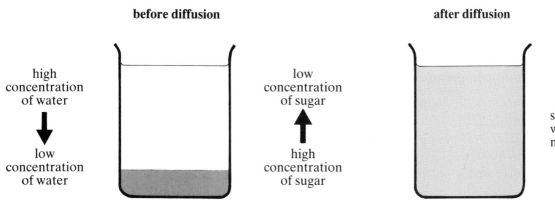

before diffusion

high concentration of water

low concentration of water

low concentration of sugar

high concentration of sugar

after diffusion

sugar and water are mixed evenly

Diffusion of gases is faster than diffusion of liquids because there are larger spaces between the fast moving molecules of a gas. Useful substances like oxygen diffuse into a cell. Waste substances like carbon dioxide diffuse out of a cell. Substances diffuse from one part of a cell to another.

Questions

1 Look at the diagram.
   (a) Where is there a high concentration of sugar molecules?
   (b) What will eventually happen to the sugar molecules?
   (c) Explain why you stir a cup of coffee after putting in a lump of sugar.

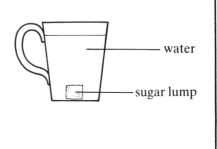

water

sugar lump

2 What would happen if you put a drop of ink into a jar of water?

3 Would diffusion of a liquid be faster or slower than diffusion of a gas? Explain your answer.

4 Name one useful substance which diffuses into a cell.

5 Name one substance which diffuses out of a cell.

Membranes

Each cell is surrounded by a cell membrane. Inside the cell there are many different molecules. If you think of the cell membrane as a bag and the molecules inside the cell as shopping it should help you to understand how some molecules pass through a cell membrane.

A polythene bag would not allow oranges , plums ○ or cherries ● to pass through. A polythene bag is therefore **impermeable**.

A string bag would not allow oranges to pass through but plums and cherries could. The size of the hole 'selects' which things pass through and which things do not. A string bag is **selectively permeable**.

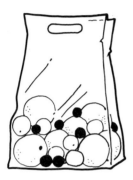

An experiment using a selectively permeable membrane

A cell membrane is selectively permeable. In this experiment a selectively permeable membrane is used to make two model cells. Each model cell was weighed before it was put into a beaker.

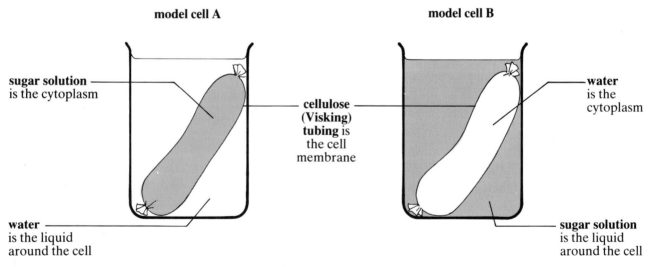

model cell A

model cell B

sugar solution is the cytoplasm

cellulose (Visking) tubing is the cell membrane

water is the cytoplasm

water is the liquid around the cell

sugar solution is the liquid around the cell

Results	model cell A	model cell B
Weight of model cell before it was put in a beaker	9·5 grams	9·5 grams
Weight of model cell after 30 minutes	9·8 grams	8·5 grams

This diagram explains what happened to model cell A.

○ water molecules could go through the membrane

● sugar molecules were too large to go through the membrane

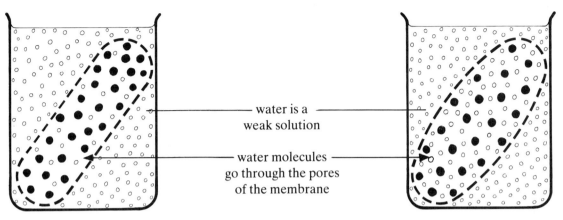

water is a
weak solution

water molecules
go through the pores
of the membrane

The sugar solution inside model cell A
is a strong solution.

Model cell A became larger as more water
molecules passed through the membrane.

Osmosis

In the last experiment model cell A gained weight because water went into
the cell by **osmosis**. Osmosis takes place when two solutions are separated by
a selectively permeable membrane. The **water** molecules from a **weak
solution** go through a **selectively permeable membrane** into a **strong solution**.

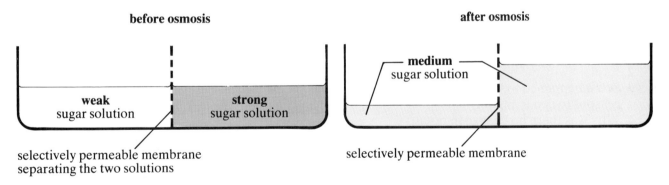

before osmosis

after osmosis

weak
sugar solution

strong
sugar solution

medium
sugar solution

selectively permeable membrane
separating the two solutions

selectively permeable membrane

Questions

1 Is a cell membrane **(a)** impermeable or **(b)** selectively permeable?
2 What is the difference between an impermeable and a selectively
permeable membrane?
3 Read 'an experiment using a selectively permeable membrane'.
Explain why model cell B lost weight. You may draw a diagram to help
you.

Osmosis in living cells

Water moves from cell to cell by osmosis. In this experiment potato cells, with selectively permeable membranes, separate the sugar from the water in the dish.

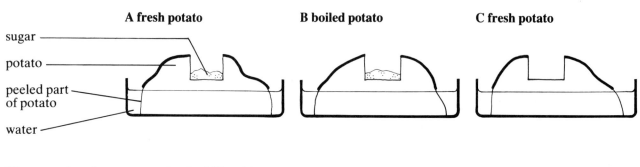

A fresh potato **B boiled potato** **C fresh potato**

sugar

potato

peeled part of potato

water

The next day the potatoes looked like this:

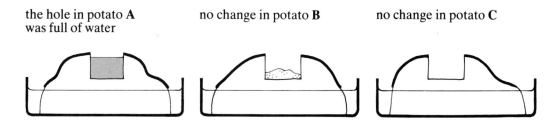

the hole in potato **A** was full of water   no change in potato **B**   no change in potato **C**

Questions

Read 'Osmosis in living cells' before answering these questions:
1 Do potato cells have selectively permeable membranes?
2 Explain why the hole in potato A filled with water.
3 What happened to the cells of potato B when it was boiled? Give reasons for your answer.
4 Why was there no change in potato C?

Summary

Living things are made up of cells. Each cell is a mass of cytoplasm containing a nucleus. The nucleus and cytoplasm are surrounded by a cell membrane.

Everything is made up of molecules which are always moving. Molecules diffuse from a high concentration to a low concentration until all the molecules are mixed evenly. Substances move from one part of a cell to another by diffusion.

Osmosis takes place when two solutions are separated by a selectively permeable membrane. Small water molecules can go through the pores of a selectively permeable membrane, but larger molecules cannot. Water from a weak solution goes through a selectively permeable membrane into a strong solution until both solutions are the same strength.

Key words

cell	A cell is made up of a nucleus, cytoplasm and a cell membrane
cell wall	Firm layer on the outside of a plant cell. Made of cellulose
cell membrane	Very thin layer on the outside of a cell
chlorophyll	Green substance found in plant cells
chloroplast	Part of a plant cell containing chlorophyll
cytoplasm	The jelly-like substance surrounding the nucleus of a cell
diffusion	Molecules move from a high concentration to a low concentration until they are mixed evenly
nucleus	Controls the activities of a cell
osmosis	Water from a weak solution goes through a selectively permeable membrane into a strong solution
selectively permeable membrane	Small water molecules can go through a selectively permeable membrane but larger molecules cannot
vacuole	Space in the cytoplasm of a cell. Vacuoles in plant cells are filled with cell sap

Questions

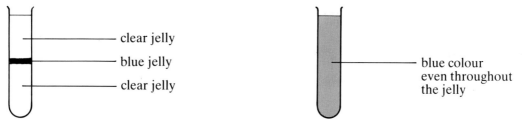

1 Look at this animal cell.
Name the parts labelled A B and C.

2 Look at this plant cell.
Name the parts labelled P Q R S and T.

3 What is a vacuole?
4 What is a cell?
5 Give two ways in which animal and plant cells are similar.
6 Some flowers have a strong scent. How would the scent of a flower reach your nose.

7 A test tube was set up like this: A week later it looked like this:

clear jelly
blue jelly
clear jelly

blue colour
even throughout
the jelly

(a) Where was the blue colour concentrated at the beginning of the experiment?
(b) Does this experiment show diffusion or osmosis?
(c) Why did it take a week for the blue colour to spread to all parts of the tube?

8 Look carefully at this experiment to show diffusion of a gas.

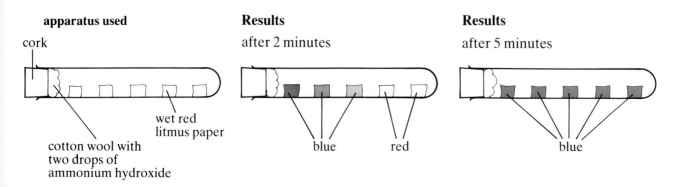

apparatus used

cork

wet red
litmus paper

cotton wool with
two drops of
ammonium hydroxide

Results
after 2 minutes

blue   red

Results
after 5 minutes

blue

(a) Why did the litmus paper turn blue? (clue – ammonium hydroxide turns red litmus blue)
(b) Would diffusion be faster or slower if one drop of ammonium hydroxide had been used? Give reasons for your answer.
(c) Why is the diffusion of gases faster than diffusion of liquids?

9 A model cell was made like this:

After 30 minutes it looked like this:

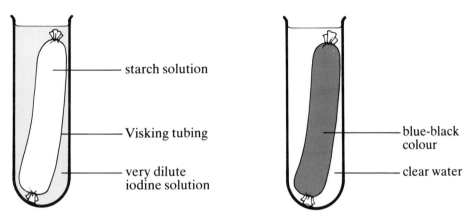

When starch solution and iodine solution are mixed together, the solution turns blue-black. Now you can answer these questions:
(a) What has happened to the iodine molecules?
(b) Which molecules (starch or iodine) did not pass through the membrane?
(c) Is Visking tubing impermeable or selectively permeable?

10 Two chips were cut from a potato. Each chip was 50 mm long. One chip was put into water. The other chip was put into a strong sugar solution.
    The next day each chip was taken out of the liquid and dried with blotting paper. This diagram shows what the potato chips looked like.

chip A   chip B

Use a ruler to measure the length of each chip.
(a) Which chip had been in water? Give reasons for your answer.
(b) Which chip had been in a strong sugar solution? Give reasons for your answer.

3: How cells work

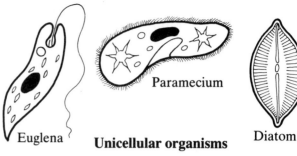

Some plants and animals have only one cell. They are called **unicellular organisms**. All the life processes take place in the one cell of a unicellular organism.

Euglena   Paramecium   Diatom

Unicellular organisms

A unicellular organism – Amoeba

Amoeba is a unicellular animal which lives in ponds. It moves, feeds, respires, excretes, is sensitive, reproduces and grows. The different parts of an Amoeba can be seen by using a microscope.

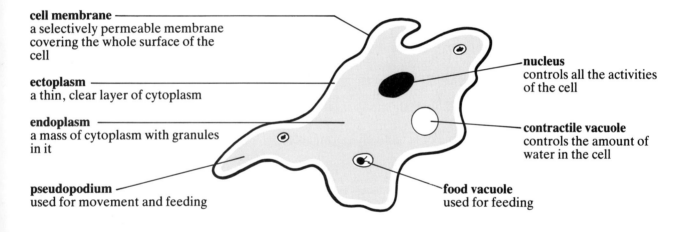

cell membrane
a selectively permeable membrane covering the whole surface of the cell

ectoplasm
a thin, clear layer of cytoplasm

endoplasm
a mass of cytoplasm with granules in it

pseudopodium
used for movement and feeding

nucleus
controls all the activities of the cell

contractile vacuole
controls the amount of water in the cell

food vacuole
used for feeding

Movement

An Amoeba changes shape when it moves.

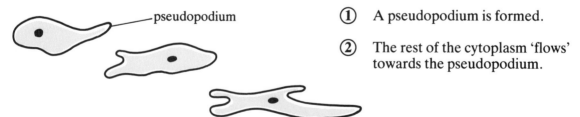

pseudopodium

① A pseudopodium is formed.

② The rest of the cytoplasm 'flows' towards the pseudopodium.

In this way the Amoeba 'flows' from one place to another. This is known as **amoeboid movement**.

Feeding

Amoeba feeds on bacteria, algae and other unicellular organisms.

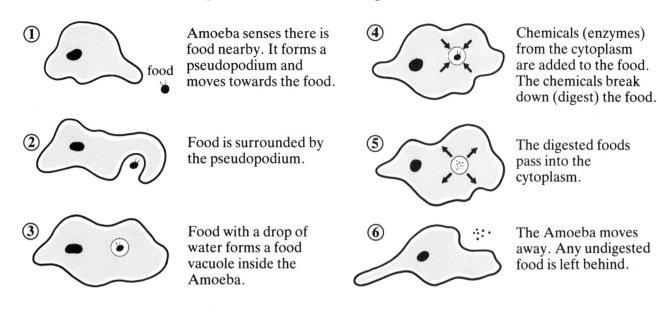

① Amoeba senses there is food nearby. It forms a pseudopodium and moves towards the food.

food

② Food is surrounded by the pseudopodium.

③ Food with a drop of water forms a food vacuole inside the Amoeba.

④ Chemicals (enzymes) from the cytoplasm are added to the food. The chemicals break down (digest) the food.

⑤ The digested foods pass into the cytoplasm.

⑥ The Amoeba moves away. Any undigested food is left behind.

Respiration

Oxygen is dissolved in the water around the Amoeba. Oxygen diffuses through the whole surface of the cell.

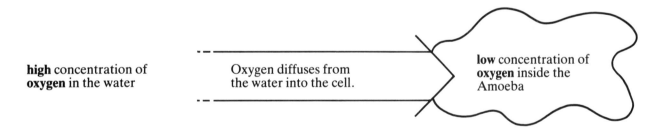

high concentration of **oxygen** in the water

Oxygen diffuses from the water into the cell.

low concentration of **oxygen** inside the Amoeba

Oxygen reacts with food in the cell to give energy which is needed for other processes.

Questions

1 The cell membrane covers the whole surface of the Amoeba.
  (a) Can the cell membrane change shape?
  (b) Can substances pass through the cell membrane? Explain your answer.

2 What is ectoplasm?
3 What is a pseudopodium used for?
4 Explain how an Amoeba feeds.
5 Explain how oxygen gets into an Amoeba.
6 Why does an Amoeba need oxygen?

Excretion

Amoeba makes waste substances which are poisonous. Waste substances must be removed (excreted) from the cell.

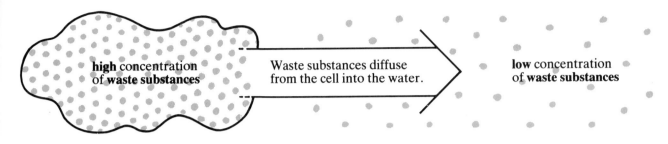

An Amoeba can control the amount of water in the cell. Water goes into the cell by osmosis.

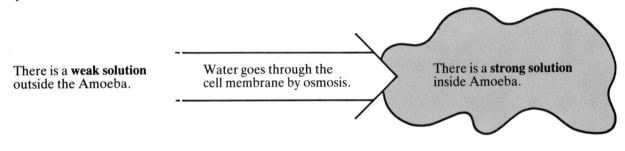

The contractile vacuole

The contractile vacuole controls the amount of water in the cell.

 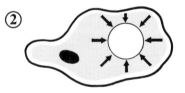 

The water goes into a **contractile vacuole**.

As more water enters, the contractile vacuole gets bigger.

When the contractile vacuole bursts water is removed from the cell. This is known as **osmoregulation**.

Sensitivity

Amoeba moves **towards food**.   Amoeba moves **away from strong light**.

Growth and reproduction

Amoeba grows until it reaches a certain size. It then reproduces by splitting into two. This process is called **binary fission**.

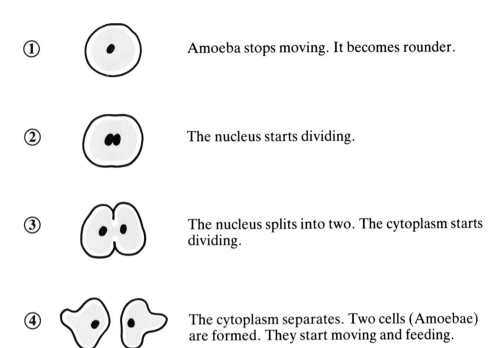

①   Amoeba stops moving. It becomes rounder.

②   The nucleus starts dividing.

③   The nucleus splits into two. The cytoplasm starts dividing.

④   The cytoplasm separates. Two cells (Amoebae) are formed. They start moving and feeding.

Questions

1 When an Amoeba respires it makes carbon dioxide. Carbon dioxide is a waste substance. How is carbon dioxide removed from the cell?
2 How would Amoeba respond to strong light?
3 How would Amoeba respond to food? Why do you think an Amoeba responds this way?
4 What is binary fission?
5 Why is the contractile vacuole important?
6 Explain how Amoeba removes water from its cell.

A simple multicellular plant – Spirogyra

All the life processes take place in the one cell of an Amoeba. Some organisms have many cells. They are called **multicellular organisms**.

Spirogyra is a multicellular plant which floats on the surface of ponds. Spirogyra is made up of many cells which all look the same. All the life processes take place in every cell. The cells of Spirogyra are joined end-to-end to make a thread or **filament**.

Filament of Spirogyra

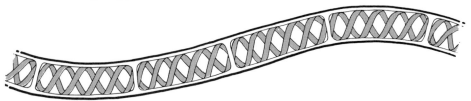

Section across one cell

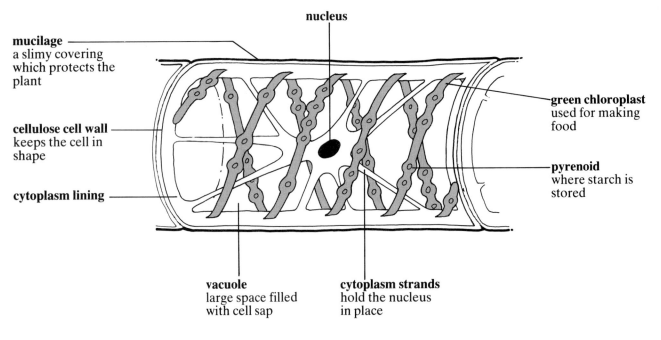

nucleus

mucilage — a slimy covering which protects the plant

cellulose cell wall keeps the cell in shape

cytoplasm lining

green chloroplast used for making food

pyrenoid where starch is stored

vacuole large space filled with cell sap

cytoplasm strands hold the nucleus in place

Questions

1 Where is Spirogyra found?
2 Is Spirogyra a unicellular organism or a multicellular organism? Give a reason for your answer.
3 What does mucilage do?
4 What colour is a chloroplast? What is a chloroplast used for?

A complicated multicellular plant – Buttercup

Spirogyra is a simple multicellular plant. All the cells look alike and do the same work. Larger and more complicated plants and animals are made up of many different cells. The work of the body is shared by all the cells. This is known as **division of labour**.

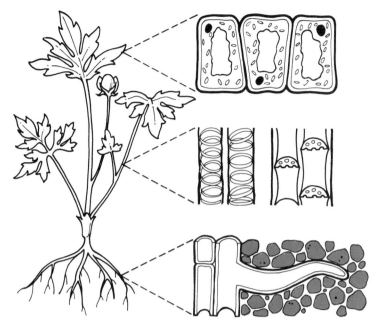

leaf cells
make food for the whole plant

stem cells
1 carry food from the leaves to the roots
2 carry water and minerals from
   the roots to the leaves

root cells
take in water and minerals
for the whole plant

A complicated multicellular animal – Man

A man is made up of many different cells. There is a division of labour between the different types of cells.

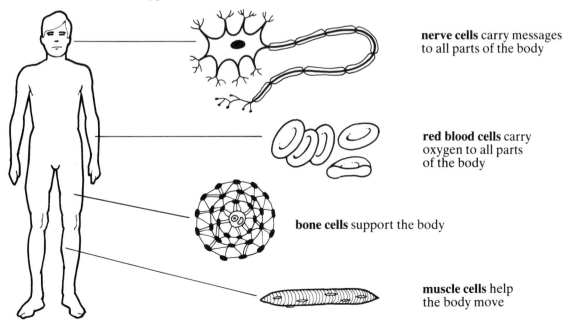

nerve cells carry messages
to all parts of the body

red blood cells carry
oxygen to all parts
of the body

bone cells support the body

muscle cells help
the body move

21

Cells, tissues and organs

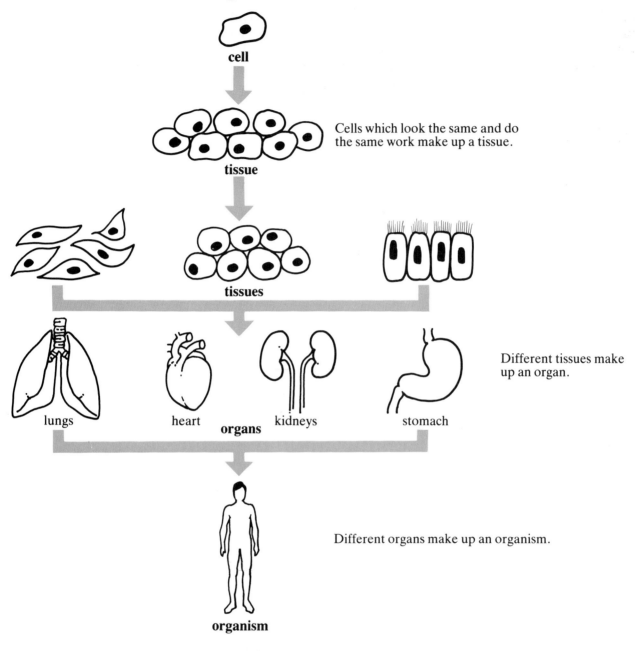

cell

Cells which look the same and do the same work make up a tissue.

tissue

tissues

Different tissues make up an organ.

lungs   heart   kidneys   stomach

organs

Different organs make up an organism.

organism

Questions

1 What work is done by the leaf cells of a plant?
2 Which cells of a plant take in water and minerals?
3 What work is done by nerve cells?
4 Which cells carry oxygen to all parts of the body?
5 Which cells help a man to move?
6 Name three organs found in the human body.

Summary

In unicellular organisms all the life processes take place in one cell. In multicellular organisms the work is shared by all the cells in the body.

Cells which do the same work are grouped together to make tissues. Different tissues are grouped together to make an organ. Different organs make up an organism.

Key words

contractile vacuole	Controls the amount of water in an Amoeba
multicellular organism	A plant or animal made of many cells
organ	A group of different tissues working together e.g. the heart, a leaf
pseudopodium	Part of an Amoeba used for moving and feeding
tissue	A group of cells which look alike and do the same work
unicellular organism	One celled plant or animal

Questions

1 Name a unicellular animal.
2 Name two multicellular organisms.
3 Draw an Amoeba and label the nucleus, cytoplasm and cell membrane.
4 What is a pseudopodium?
5 Name an organism which has a food vacuole. How is a food vacuole formed?
6 In your own words, describe how an Amoeba reproduces.

7 Here are three cells (not drawn to scale). They are all found in the human body.

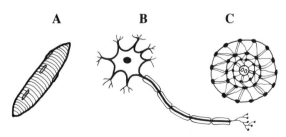

   A   B   C

(a) Name each type of cell. (b) What is the function of each type of cell?

8 Explain what is meant by a tissue.
9 Explain what is meant by an organ. Name one plant organ and one animal organ.
10 What is meant by division of labour?

11 In diagrams cells look flat. These sectional views show that cells are not flat.
(a) Letters A to E show parts of an Amoeba. Name each part.

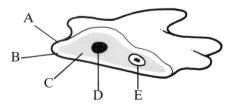

(b) Letters F and G show parts of Spirogyra. Name each part.

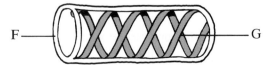

4: Energy

When you opened this book you used energy. As you read this page you use energy. Everything you do needs energy. We get our energy from the food we eat.

All our food comes from green plants. The next diagram explains how the food we eat comes from green plants.

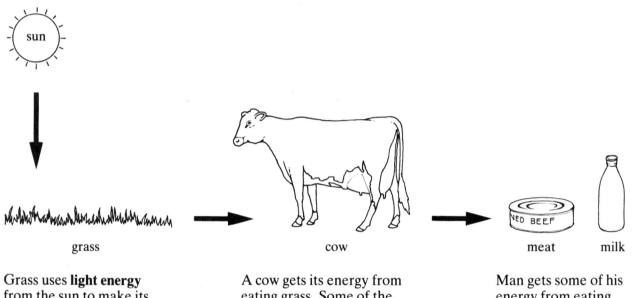

| grass | cow | meat   milk |

Grass uses **light energy** from the sun to make its own food.

A cow gets its energy from eating grass. Some of the energy is used to make meat and milk.

Man gets some of his energy from eating meat and drinking milk.

A food chain

All animals get their energy from eating plants or other animals which have eaten plants. Energy passes from green plants to animals along a **food chain**.

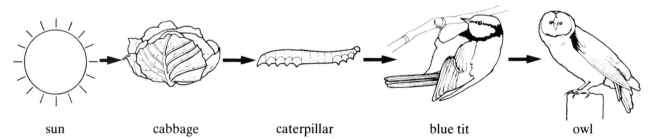

| sun | cabbage | caterpillar | blue tit | owl |

Green plants are called **producers** because they make (produce) their own food.

Animals are called **consumers** because they eat (consume) plants or other animals.

Passing on energy

Some energy is wasted as it passes along a food chain. A field of cabbages would be enough food for many small caterpillars. The caterpillars would be enough food for a few blue tits. The blue tits would be enough food for one owl. If all the organisms in each link of the food chain were piled on top of each other they would make a pyramid.

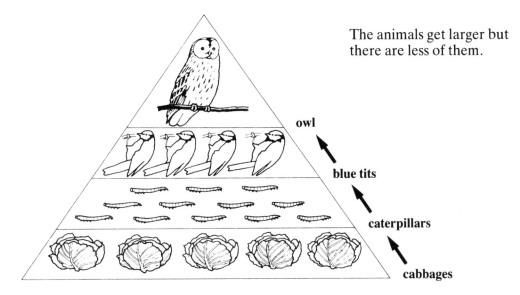

The animals get larger but there are less of them.

owl

blue tits

caterpillars

cabbages

Questions

1 Why are green plants called producers?
2 Name two producers
3 Explain how man depends on green plants for his food.
4 Why are animals called consumers?

Storing food

During the summer some plants make more food than they need. The extra food goes into special organs which store the food. The stored food keeps the plant alive during the winter and provides the energy needed to grow new leaves next spring.

A carrot stores food in a **tap root**.

A potato stores food in a **tuber**.

An onion stores food in a **bulb**.

A crocus stores food in a **corm**.

How animals survive the winter

Animals use up a lot of energy moving, growing and looking for food. In the winter there is not enough food for some animals to get all the energy they need. They must store food during the summer or save energy by resting during the winter.

Food storage

Bees make honey in the summer. They feed on the stored honey during the winter.

A squirrel stores food under the ground so that it will have enough to eat during the winter.

Hibernation

During the summer a hedgehog eats more food than it needs. The food is changed into fat and stored under the skin and round the kidneys, heart and liver.

A hedgehog **hibernates** during the winter. It stops moving and its heart beat gets slower so that it uses very little energy. All its energy comes from the stored fat.

Migration

Birds which cannot get enough food during the winter fly or **migrate** to another country. Before migrating a bird stores fat so that it will have enough energy for the long flight.

A swallow migrates to Africa.

Questions

1 Name three plant organs which store food.
2 Name two animals which store food so that they will have enough to eat during the winter.
3 How does a hedgehog prepare for hibernation?
4 Name a bird which migrates to a warmer country for the winter.

Summary

Life on earth depends on green plants because they use the sun's energy to make their own food. Animals get their energy from eating green plants or other animals which have eaten green plants. Animals get the energy they need for respiring, feeding and growing from the food they eat.

Energy passes along a food chain. If all the plants and animals in a food chain are piled on top of each other they make a pyramid.

Some plants store food in bulbs, tubers or corms to help them survive during the winter. Some animals save energy by resting or hibernating during the winter. Some birds migrate to other parts of the world for the winter.

Key words

consumers   Animals which get their food from green plants

hibernation   A long, deep sleep in winter

migration   Birds fly (migrate) to another country to find food

producers   Green plants

Questions

1 Why do some plants store food?
2 How does a squirrel make sure that it has enough food for the winter?
3 Where do bees get their energy from in the winter?
4 Why do some animals hibernate during the winter?
5 Explain why swallows migrate in the autumn.
6 How does a swallow prepare for migration?

7 Complete this table to show where the following plants store food:

Plant	Storage organ
carrot	root
onion	
crocus	
potato	

8 Draw a simple diagram to explain how a dinner of beef, cabbage and potatoes is made by green plants.

5: Photosynthesis

Green plants are called **producers** because they make their own food. They make their food by **photosynthesis**. Most photosynthesis takes place in the leaves of a plant. Chlorophyll (the green colour) in the leaves traps light energy from the sun. The energy is used to change water and carbon dioxide into sugar and oxygen.

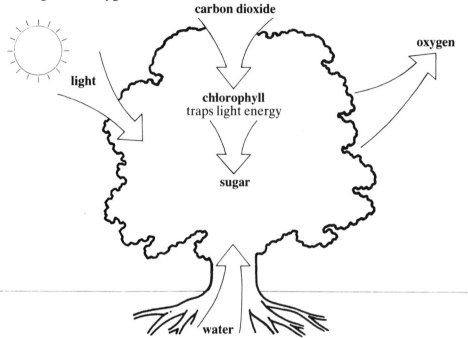

This equation shows what happens.

$$\text{water} \; + \; \text{carbon dioxide} \; \xrightarrow[\text{chlorophyll}]{\text{sunlight energy}} \; \text{sugar} \; + \; \text{oxygen}$$

This is another way of writing the equation.

$$6H_2O \; + \; 6CO_2 \; \xrightarrow[\text{chlorophyll}]{\text{sunlight energy}} \; C_6H_{12}O_6 \; + \; 6O_2$$

Questions

1 What is chlorophyll?
2 What does chlorophyll do?

3 How does a green plant make sugar?
4 Name the gas made during photosynthesis.

28

The parts of a leaf

Photosynthesis takes place in the leaves of a green plant. Some plants turn
their leaves towards the sun so that they have enough light for
photosynthesis. Other plants spread their leaves out so that they do not
overlap or shade each other. The leaves on some trees are spread out like a
large umbrella so that each leaf gets as much light as possible.

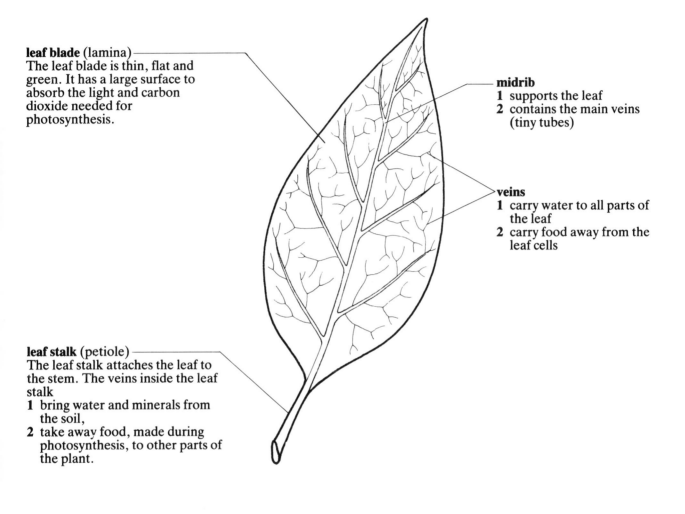

leaf blade (lamina)
The leaf blade is thin, flat and
green. It has a large surface to
absorb the light and carbon
dioxide needed for
photosynthesis.

midrib
1 supports the leaf
2 contains the main veins
 (tiny tubes)

veins
1 carry water to all parts of
 the leaf
2 carry food away from the
 leaf cells

leaf stalk (petiole)
The leaf stalk attaches the leaf to
the stem. The veins inside the leaf
stalk
1 bring water and minerals from
 the soil,
2 take away food, made during
 photosynthesis, to other parts of
 the plant.

Questions

1 What is the midrib?
2 What is the function of the midrib?
3 Why does the leaf blade have a large, thin
 surface?
4 Why is the leaf stalk important?
5 Why are veins important?
6 Explain why a plant may twist the leaf
 blade towards the sun.

The structure of a leaf

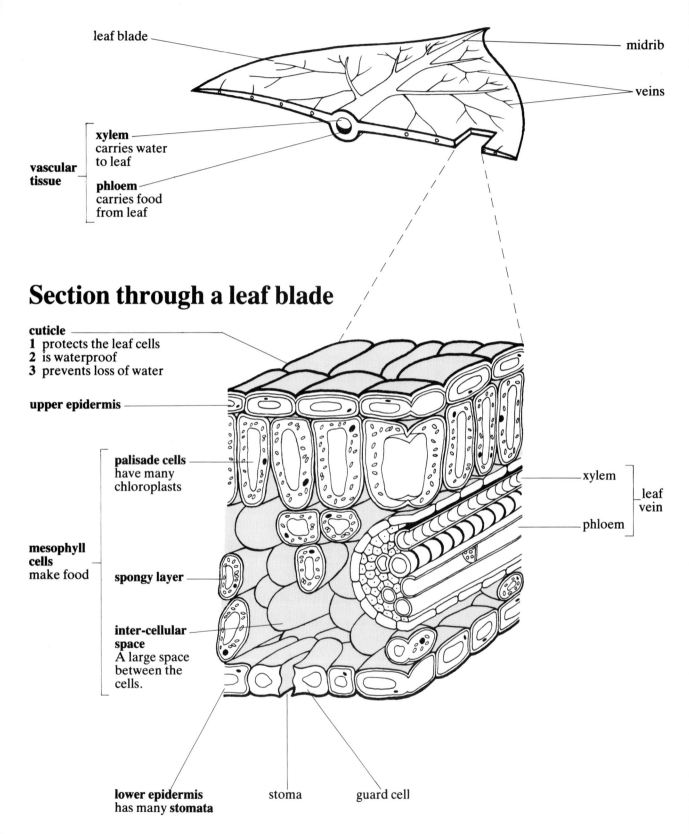

leaf blade

midrib

veins

vascular tissue

xylem
carries water
to leaf

phloem
carries food
from leaf

Section through a leaf blade

cuticle
1 protects the leaf cells
2 is waterproof
3 prevents loss of water

upper epidermis

palisade cells
have many
chloroplasts

mesophyll cells
make food

spongy layer

inter-cellular space
A large space
between the
cells.

xylem

phloem

leaf vein

lower epidermis
has many **stomata**

stoma

guard cell

Stomata

There are many stomata on the lower epidermis of a leaf. Each stoma is surrounded by guard cells.

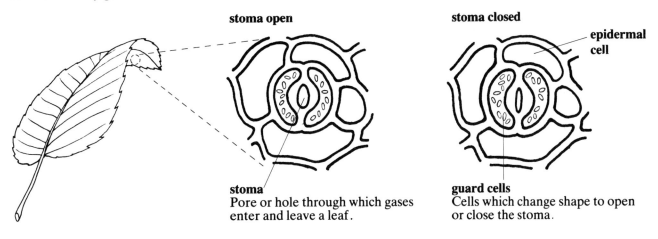

stoma open

stoma closed

epidermal cell

stoma
Pore or hole through which gases enter and leave a leaf.

guard cells
Cells which change shape to open or close the stoma.

How a plant uses the sugar made by photosynthesis

The sugar made by photosynthesis can be used by the plant or changed into starch. The starch can be stored in the leaf cells and other parts of the plant until it is needed.

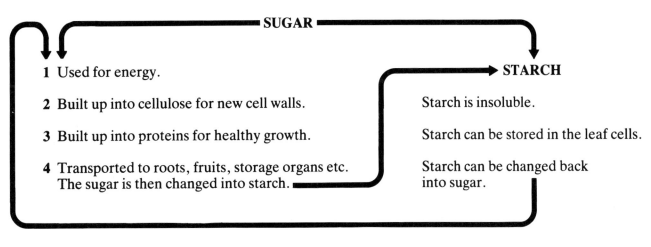

SUGAR

1 Used for energy.

2 Built up into cellulose for new cell walls.

3 Built up into proteins for healthy growth.

4 Transported to roots, fruits, storage organs etc. The sugar is then changed into starch.

STARCH

Starch is insoluble.

Starch can be stored in the leaf cells.

Starch can be changed back into sugar.

Questions

1 Give two functions of the cuticle of a leaf.
2 Name the cells which make food.
3 How is water carried to every cell in the leaf?
4 Which vessels carry sugar away from the leaf?

5 What are stomata?
6 How does each stoma open and close?
7 How does a plant use the sugar made by photosynthesis?
8 Where does a plant store starch?

How to test a leaf for starch

In this experiment two leaves are being tested for starch.

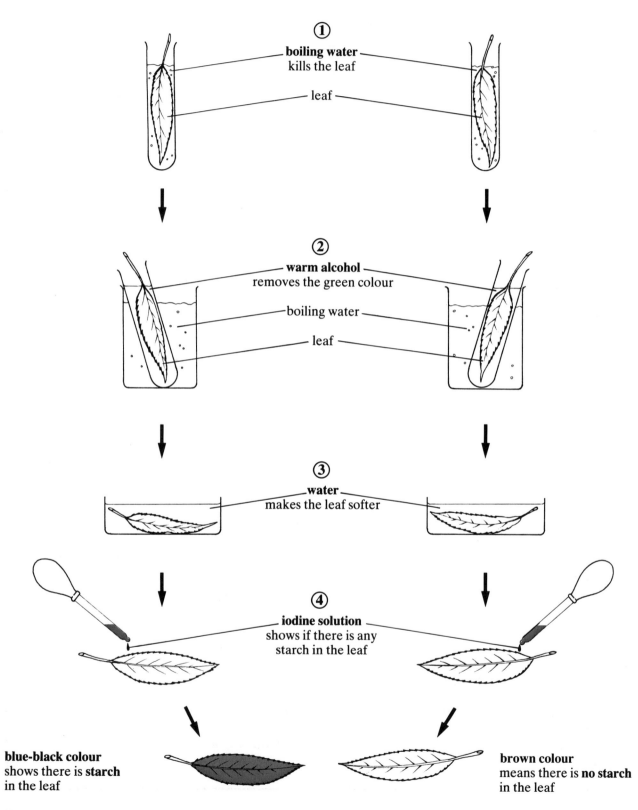

① **boiling water** kills the leaf

leaf

② **warm alcohol** removes the green colour

boiling water

leaf

③ **water** makes the leaf softer

④ **iodine solution** shows if there is any starch in the leaf

blue-black colour shows there is **starch** in the leaf

brown colour means there is **no starch** in the leaf

The minerals needed by plants

A green plant needs water and carbon dioxide to make sugar. A plant uses the chemicals in sugar to make starch, fats and oils. A plant needs some other chemicals to make proteins and to stay healthy. These chemicals are called **minerals**. Nitrogen, magnesium and iron are three of the minerals found in soil. These minerals are dissolved in the water which is taken in by the roots.

This plant has been grown in soil with **all the minerals** needed for normal healthy growth.

Nitrogen is needed for making proteins. This plant has been grown in soil **without nitrogen**.

Magnesium is needed for making chlorophyll. This plant has been grown in soil **without magnesium**.

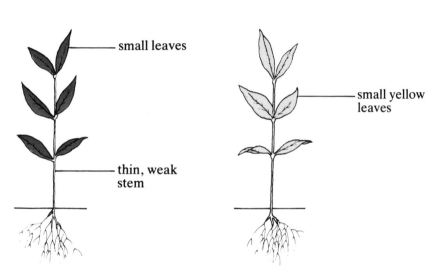

small leaves

thin, weak stem

small yellow leaves

Questions

1 Copy these sentences and fill in the missing words:

How to test a leaf for starch.
First kill the leaf in .... Soak the leaf in warm.................
to remove the green colour. Soften the leaf by placing it in ....
Add............solution to the leaf.

2 What happens when iodine solution is added to starch?
3 Name two minerals needed by a plant.
4 Explain why a plant needs the minerals you have named.
5 How does a plant take in the minerals it needs?

Summary

Green plants make their own food by photosynthesis. They use light energy from the sun to make sugar from water and carbon dioxide. Photosynthesis takes place in the leaves and other green parts of a plant.

Sugar, made by photosynthesis, can be used to supply energy or can be changed into starch for storage. Some food is stored in the stem or roots until it is needed by the plant. The chemicals in sugar can be used to make starch, fats and oils. Plants also need minerals from the soil to make proteins and to stay healthy.

Key words

chlorophyll   The green colour in plants. Used to trap light energy from the sun.

phloem   Carries food.

photosynthesis   How green plants make their food.

stomata   Pores or holes in a leaf. One stoma. Many stomata.

xylem   Carries water.

Questions

1 Copy and complete this paragraph by using the following words:–

starch   photosynthesis   sugar   water   chlorophyll   energy

Green plants make their own food. The green............ in the chloroplasts of the leaf cells traps the sun's .... They use carbon dioxide and .... to make .... This process is called .... If there is .... in the leaf cells it shows that photosynthesis has taken place.

2 A plant with variegated leaves (part green and part yellow) was left in the sun for five hours. Then one leaf was taken off the plant and tested for starch. Only part of the leaf had starch.

a variegated leaf

green   yellow

(a) Describe how you would test the leaf for starch.

(b) In which part of the leaf (green or yellow) would you expect to find starch? Give reasons for your answer.

3 Three pot plants were kept in a dark cupboard before being used for this experiment. There was no starch in the leaves at the start of the experiment. Each plant was left under a bell jar for ten hours.

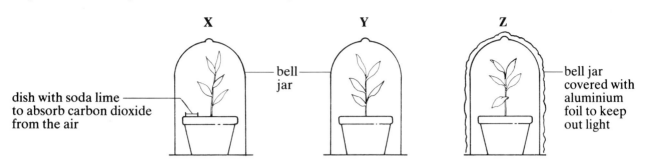

(a) Explain why there was no starch in the leaves at the start of the experiment.

(b) Why was bell jar Z covered with aluminium foil?

(c) Which chemical was used to absorb carbon dioxide from the air?

(d) After six hours a leaf was taken from each plant. Each leaf was tested to see if there was any starch in it.
Copy this table and fill in the results you would expect to get from this test.

Leaf from plant	Colour after testing with iodine solution
X	
Y	
Z	

(e) Explain why you would expect these results.

4 These diagrams show an experiment on photosynthesis. At the start of the experiment each test tube was full of water.

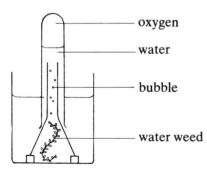

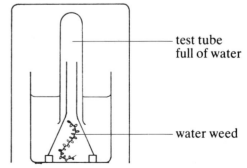

(a) Where do you think the bubbles in tube A came from?

(b) Which gas would you expect to be in the bubbles? Give reasons for your answer.

(c) Why was tube B full of water at the end of the experiment?

6: Food and diet

Plants trap the light energy from the sun and store it in the food they make. The food we eat comes from green plants or from animals which have eaten green plants. Food gives us the energy and the other materials needed for growth and repair of the body.

Types of food

There are three types of food – **carbohydrates, fats** and **proteins**. We need carbohydrates and fats for energy. We need proteins for growth and repair.

Carbohydrates (starch and sugar)

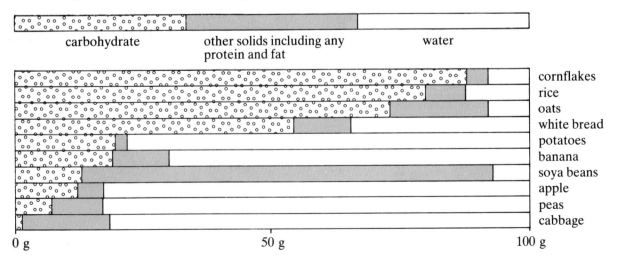

Fats

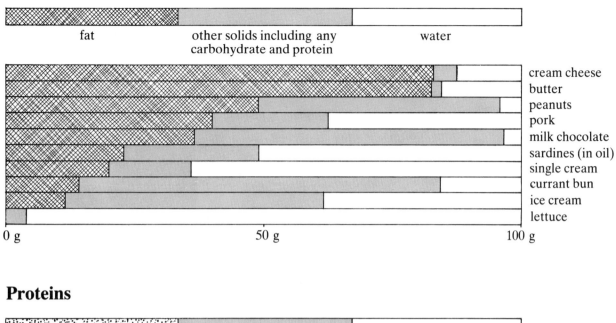

Proteins

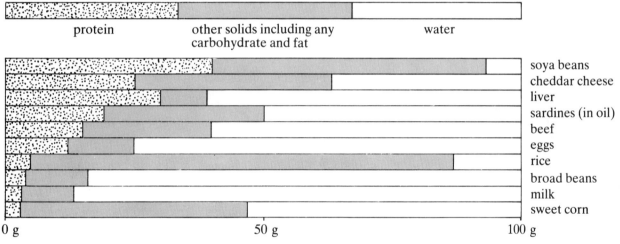

Questions

On the graphs, each strip equals 100 g of food. The graphs show that food is made of solids and water.

1 Using a ruler, measure the strips to find out which food has the most **(a)** carbohydrate **(b)** fat **(c)** protein **(d)** water.
2 Which contains more carbohydrate, 100 g of potatoes or 100 g of white bread?
3 Which contains more protein, 50 g of soya beans or 100 g of liver?
4 Soya beans are cheaper than meat. Why do you think soya beans are sometimes used instead of meat?
5 Which contains more water, cheddar cheese or cream cheese?

Food tests

This table shows the reagents used to test food for carbohydrates and protein.

Food tested	Type of food	Reagent used	Result
potato	starch (carbohydrate)	iodine solution	blue-black colour
glucose	sugar (carbohydrate)	Benedict's reagent	orange-red colour
egg white	protein	Millon's reagent	red colour

fat leaves a grease spot on paper

grease spot

Questions

1 Which reagent would you use to test for sugar?
2 Which type of food would you test with Millon's reagent?
3 If a grease spot is left on a piece of paper does it show that the food contained (**a**) starch (**b**) sugar or (**c**) fat?

Other important food substances

Carbohydrates, fats and proteins give us energy and the materials needed for growth and repair of the body. We also need very small amounts of **vitamins** and **minerals** to stay healthy. The food we eat contains very small amounts of vitamins and minerals.

Vitamins

Vitamin	Foods rich in vitamin	Why the body needs the vitamin	What happens if the body does not get enough of the vitamin
A	milk, butter, fresh green vegetables, carrots	needed for: healthy skin and good vision in dim light	The person cannot see very well in dim light. The person suffers from **night blindness**.
B (group of vitamins)	yeast, liver, wholemeal bread	needed for: healthy nerves	The nervous system is affected and the muscles become weak. the person suffers from **beri-beri**.
C	oranges, grapefruit, lemons, fresh green vegetables	needed for: healthy skin and blood vessels	The gums and nose bleed. There is bleeding inside the body and wounds heal more slowly. The person suffers from **scurvy**.
D	milk, butter, eggs, fish liver oils	needed for: strong bones and teeth	The bones become soft and bend or break easily. The person suffers from **rickets**.

Minerals

Mineral	Foods rich in the mineral	Why the body needs the mineral
calcium	milk, eggs	used for making bones and teeth
iodine	fish, salt	It is used by the thyroid gland. The thyroid gland controls growth.
iron	liver, spinach	used for making red blood cells
phosphorus	meat, milk, fish	used for making bones and teeth

Water

Water is not a food but without it a person would die in a few days. Water cannot be stored in the body. When water is lost from the body more water must be taken in.

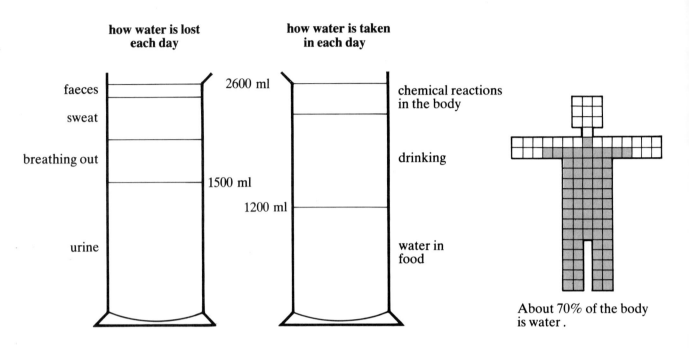

how water is lost each day

faeces

sweat

breathing out

urine

2600 ml

1500 ml

how water is taken in each day

chemical reactions in the body

drinking

1200 ml

water in food

About 70% of the body is water.

Questions

1 What would happen if you did not get enough vitamin A?
2 Name three foods which are rich in vitamin C.
3 Which two vitamins are found in milk?
4 Why does a person need vitamin D?
5 What is beri-beri?
6 Which mineral is needed for making red blood cells?
7 What is calcium used for?
8 Which two minerals are needed for strong teeth and bones?
9 Use 100 squares to draw a 'man'. Using different colours, shade in:
   (a) 8 squares for the amount of water in the blood
   (b) 18 squares for the amount of water between the cells of the body
   (c) 45 squares for the amount of water in the cells of the body.
10 How does the body lose water?
11 How does the body get most of its water each day?

Balanced diet

A person needs carbohydrates, fats, proteins, vitamins, minerals and water to stay healthy. A **balanced diet** has the right amounts of all these things.

There is a balance between the food we eat and the energy we use up. The amount of food needed depends on the **age, sex** and **work done** by the person.

Age: a man needs more food than a baby.

Sex: a man needs more food than a woman.

Work: a footballer needs more food than a man reading a book.

The amount of energy used

The **energy value** of food is measured in **kilojoules (kJ)**. This graph shows the amount of energy used in a day.

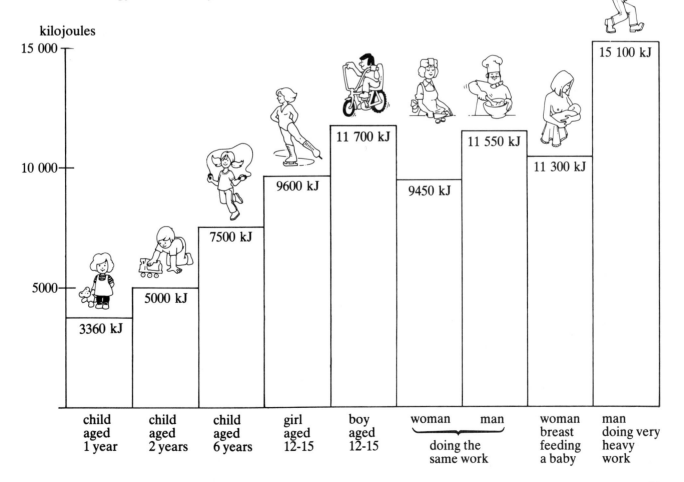

child aged 1 year	child aged 2 years	child aged 6 years	girl aged 12-15	boy aged 12-15	woman	man	woman breast feeding a baby	man doing very heavy work
3360 kJ	5000 kJ	7500 kJ	9600 kJ	11 700 kJ	9450 kJ	11 550 kJ	11 300 kJ	15 100 kJ

woman — man doing the same work

Malnutrition

Millions of people die each year because they do not have enough food. Many more people live on a diet which does not have all the things needed to stay healthy. They suffer from a disease called **malnutrition**. People suffering from malnutrition are often underweight, unhealthy and too weak to work.

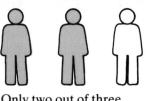

Only two out of three people have enough food.

Pellagra

Many children in Central America suffer from **pellagra** because they do not have enough **vitamin B** in their diet.

Kwashiorkor

Many people in deprived areas of the world do not have enough money to buy protein foods such as meat and fish. Many children in these areas suffer from **kwashiorkor**.

Overweight

Some wealthy people eat too much of the wrong kind of food. These people are often unhealthy because they are overweight. Although they eat a lot of food they may not get enough minerals and vitamins to stay healthy.

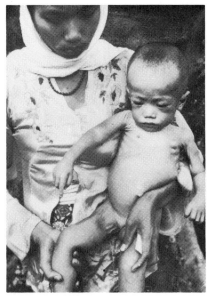

child suffering from kwashiorkor

overweight child

Questions

1 Name the things which make up a balanced diet.
2 Each person needs a different amount of energy. Why is this so?
3 A baby 1 year old needs 3360 kJ a day. How many more kilojoules a day does a child of 2 years need?
4 (a) In which part of the world are you likely to find children suffering from pellagra?
   (b) Which vitamin is missing from their diet?
5 What type of food is missing from the diet of a child suffering from kwashiorkor?

Food preservation

Food soon goes bad because there are bacteria and fungi (**micro-organisms**) which live and grow on food. When micro-organisms feed they make waste substances. Some of the waste substances are poisonous, others make the food smell and taste bad.

Food can be **preserved** by killing the micro-organisms or by making sure that they cannot live on the food.

Micro-organisms cannot live on **dried food**. Once the food has been dried it must be kept in a dry place.

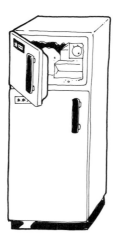

Food stays fresh if it is kept in a **refrigerator** or **freezer** because micro-organisms cannot reproduce at low temperatures.

The water is drawn out of micro-organisms if they are surrounded by **sugar** or **salt** (brine). This kills the micro-organisms.

Micro-organisms are killed when food is boiled. The food is then **sealed in a tin** to keep out the air (there are micro-organisms in the air).

Micro-organisms are killed by **vinegar, sodium nitrate** and **other chemicals**.

Summary

To stay healthy a person must have a balanced diet. A balanced diet has the right amount of carbohydrates, fats, proteins, vitamins, minerals and water. Water is not a food but it is essential for life.

Millions of people die each year because they do not get enough food. Many more suffer from malnutrition because they do not get enough food of the right kind. The amount of food needed by people depends on their age, their sex and the type of work they do. There is a balance between the food eaten and the energy used up. If the balance between the food eaten and the energy used up changes the body weight changes.

Key words

balanced diet   A diet with the right amount of carbohydrates, fats, proteins, vitamins and minerals

carbohydrates   Sugary and starchy foods. Used for energy

malnutrition   A disease caused by a very unbalanced diet

micro-organisms   Bacteria and fungi which are too small to be seen without a microscope

proteins   Meat, fish and eggs. Used for growth and repair of tissues

vitamins   Chemicals needed in small amounts to stay healthy

Questions

1 Here is a list of food

milk, ice cream, butter, bread, cabbage, beef, lettuce, apple

(a) Which food has the most protein?
(b) Which food has the most carbohydrate?
(c) Which food has the most fat?

2 Which foods provide the body with energy?

3 Why does a 15 year-old girl need more energy each day than a 2 year old girl?

4 Explain why a man digging the garden needs more food than a man resting in bed.

5 Make a list of the different ways of preserving food.

6 Deep freezing does not kill micro-organisms. What would happen to meat if it were taken out of a deep freeze and left on the kitchen table for several hours?

7 What is scurvy?

8 What is rickets?

9 How would you test a white powder to see if it contained

(a) starch (b) glucose or (c) protein?

10 This table shows the human population, of the world, over the past few centuries.

Population of the world	
Year	Numbers (in millions)
1550	400
1650	500
1750	750
1850	1250
1950	2500
1960	3000
1970	3600
1975	4000
2000	?

(a) Use these figures to draw a graph.
(b) Put a cross (**X**) for the expected population in the year 2000.
(c) How many people will there be in the year 2000?
(d) What could be done to stop people suffering from malnutrition?

7: How animals feed

Animals get their food by eating plants or other animals. Most animals spend a lot of time either looking for food or eating it. Animals feed in many different ways.

How insects feed

Some insects bite and chew their food. Other insects suck up liquids from plants or animals. The mouth parts of an insect are suited to the type of food eaten.

Locust

The locust uses its mouth parts for biting and chewing plants.

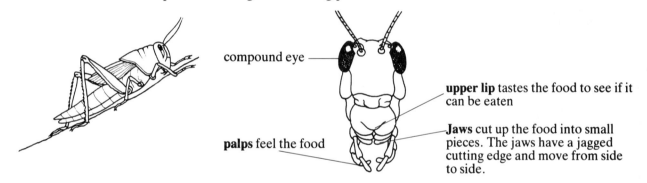

compound eye

upper lip tastes the food to see if it can be eaten

Jaws cut up the food into small pieces. The jaws have a jagged cutting edge and move from side to side.

palps feel the food

Housefly

The housefly uses its mouth parts like a sponge. It feeds on dead and decaying things like manure and dead animals as well as the food you eat.

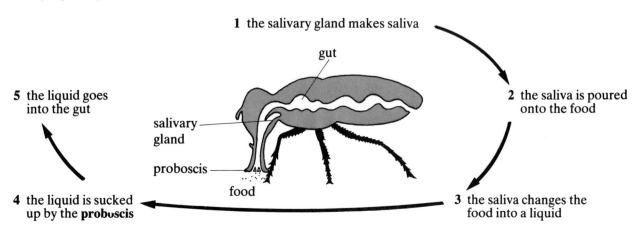

1 the salivary gland makes saliva

gut

5 the liquid goes into the gut

salivary gland

proboscis

food

4 the liquid is sucked up by the **proboscis**

2 the saliva is poured onto the food

3 the saliva changes the food into a liquid

Helpful and harmful insects

When they feed insects may damage crops, spread disease or be useful to Man.

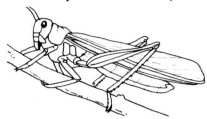

Locusts eat the crops growing in Africa.

Mosquitoes spread **malaria** when they stab the skin to suck out blood.

Houseflies spread diseases like **typhoid, cholera, dysentery** and **summer diarrhoea**.

When a **honey bee** sucks up nectar it **pollinates** the flower. It uses the nectar to make **honey**.

A **ladybird eats** the **greenflies** which damage our crops.

Questions

1 What does a locust use its mouth parts for?
2 What does a housefly feed on?
3 (a) Which part of the housefly makes saliva?
   (b) What does saliva do to the food?
4 Name two insects which spread disease when they feed.
5 How do the following insects help man
   (a) the honey bee and (b) the ladybird?

How birds feed

Birds use their beaks to pick up food. The size and shape of the beak is suited to the food eaten.

Duck

A duck swims with its beak in the water. It catches tiny insects and plants.

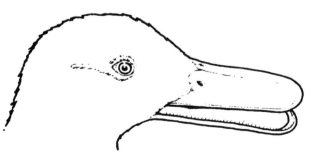

The 'bristle-like' edges trap all the food as the water and mud are squeezed out.

Eagle

An eagle uses its sharp claws to catch and kill its prey.

The strong, curved beak is used for tearing flesh.

How mammals feed

Mammals use their teeth to get food into the mouth and then cut, tear and crush it before it is swallowed. A mammal's teeth are suited to the food eaten.

mammals which eat plants are called **herbivores**

mammals which eat other animals are called **carnivores**

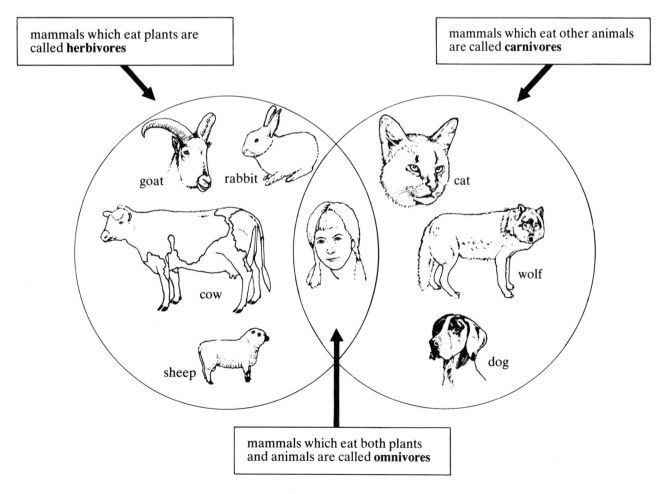

goat
rabbit
cow
sheep
cat
wolf
dog

mammals which eat both plants and animals are called **omnivores**

Types of teeth

There are four different types of teeth.

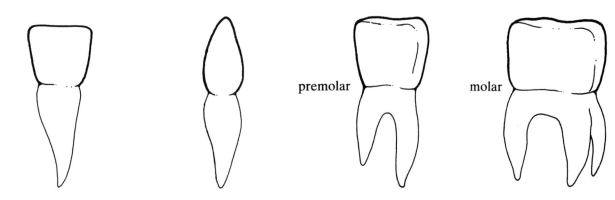

premolar

molar

incisors (I) have sharp edges to cut off the food

canines (C) are large and pointed for tearing food

premolars (PM) and **molars** (M) crush or grind up the food so that it is easy to swallow

The structure of a tooth

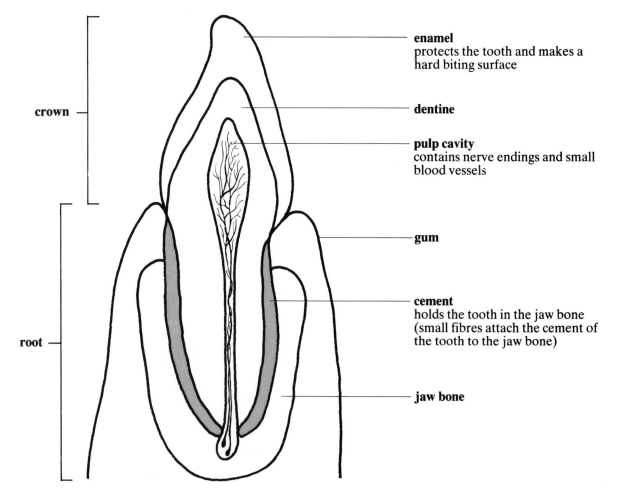

crown

root

enamel
protects the tooth and makes a hard biting surface

dentine

pulp cavity
contains nerve endings and small blood vessels

gum

cement
holds the tooth in the jaw bone (small fibres attach the cement of the tooth to the jaw bone)

jaw bone

Herbivore teeth

Herbivores have teeth which are suited to cutting and grinding grass. The teeth keep on growing so that they do not get worn away as they grind the grass.

Skull of sheep

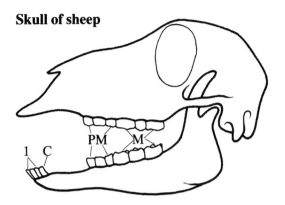

The **incisors** and **canines** are used to **grip and pull grass.**

The **premolars and molars have ridges** on them which help to **crush and grind grass.**

The lower jaw moves from side to side.

Carnivore teeth

Carnivores have teeth which are suited to killing other animals and tearing flesh. The teeth stop growing when they reach a certain size.

Skull of dog

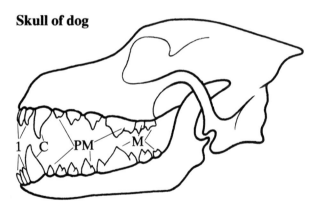

The **incisors** are used to **grip and pull flesh off bones.**

The **canines** are **long, pointed and curved.** They are used to **kill and tear pieces of flesh.**

The **premolars** work like scissors to **cut flesh and crack bones.**

The **molars crush** the food.

The lower jaw moves up and down.

Questions
1 Name the four types of teeth found in mammals.
2 What is each type of tooth used for?
3 What does a herbivore eat?
4 What does a carnivore eat?
5 Why does a dog need large, pointed and curved canine teeth?
6 Why do the premolar and molar teeth of a sheep have ridges on them?

Omnivore teeth

Man is an omnivore. His teeth are suited to eating most kinds of food. The lower jaw moves up and down. It can also move a little from side to side.

Each person has two sets of teeth in his or her lifetime. The first set of teeth is the **milk set**. The first milk tooth appears when a baby is about six months old. When a child is about five years old the milk teeth start to fall out. The milk teeth are replaced by **permanent** teeth. Once the teeth in the permanent set reach a certain size they stop growing.

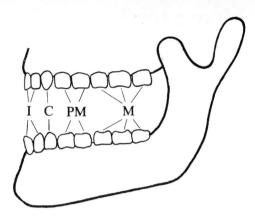

This diagram shows the permanent set of teeth.

Tooth decay

Sugar is left in the mouth after eating food. Bacteria live on the sugar and leave behind an acid which wears away the tooth enamel.

If there is a lot of food on the teeth the bacteria form a hard, white, sticky slime called **plaque**. This covers the surface of the teeth and is hard to brush off.

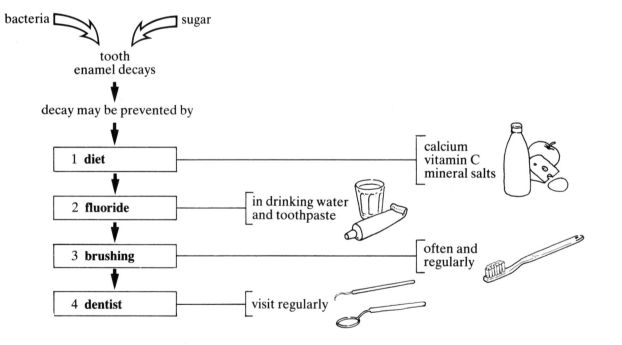

Questions

1 When does the first milk tooth appear in a child?
2 What is the second set of teeth called?
3 What causes tooth decay?
4 How can tooth decay be prevented?

Summary

The mouths and mouth parts of animals are suited to the food they eat. An insect may have mouth parts suited to sucking juices or biting and chewing plants. Birds have beaks for collecting or catching food. Mammals have teeth. The teeth of a carnivore are suited to killing animals and tearing flesh. The teeth of a herbivore are suited to cutting and grinding grass. The teeth of an omnivore are suited to eating most kinds of food.

Key words

carnivore   An animal which eats other animals

herbivore   An animal which eats plants

omnivore   An animal which eats both plants and animals

Questions

1 Why are houseflies and mosquitoes dangerous to humans when they feed?

2 (a) Give three examples of herbivores.
   (b) Give two examples of carnivores.

3 What are the main differences between the teeth of a herbivore and the teeth of a carnivore?

4 Draw a large labelled diagram to show the structure of a tooth.

5 Explain how the beaks of a duck and an eagle are suited to the food they eat.

6 When this table is completed it will show the number and type of teeth a man has in his milk set and his permanent set. Copy and complete the table. The milk set is done for you. You should be able to work out the number in the permanent set from the diagram on page 50.

		number of teeth on one side				total number in set
		I	C	PM	M	
milk set	upper jaw	2	1	2	0	20
	lower jaw	2	1	2	0	
permanent set	upper jaw					
	lower jaw					

8: Digestion

The food we eat must be broken down before it can be used by the body. The carbohydrates, fats and proteins which we eat are made up of large molecules. These large molecules are broken down into smaller molecules by **digestion**.

Digestion of carbohydrates

Large carbohydrate molecules (sugar and starch) must be broken down into smaller glucose molecules before they can be used by the body.

A **starch molecule** is made up of many glucose molecules.

digestion breaks down a
carbohydrate molecule

glucose molecules

Digestion of fats

Large fat molecules must be broken down into smaller fatty acid and glycerol molecules before they can be used by the body.

A **fat molecule** is made up of fatty acid and glycerol molecules.

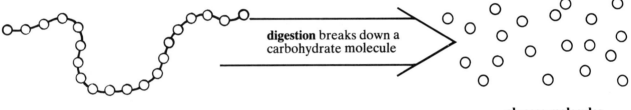

glycerol

fatty acid 1

fatty acid 2

fatty acid 3

digestion breaks down
a fat molecule

fatty acid 3

glycerol

fatty acid 1

fatty acid 2

fatty acid and **glycerol molecules**

Digestion of proteins

Large protein molecules must be broken down into smaller amino acid molecules before they can be used by the body.

A protein molecule is made up of many different amino acids.

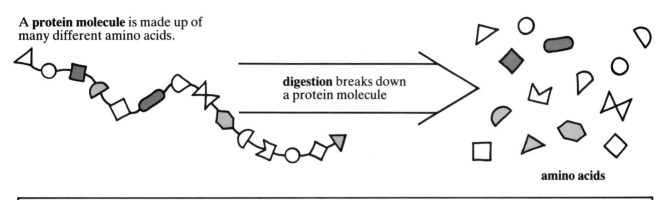

digestion breaks down
a protein molecule

amino acids

Questions

1 What does digestion do?
2 Which type of food is made up of glucose molecules?
3 Name the smaller molecules which make up a fat molecule.
4 What is a protein molecule made of?

Enzymes

Large food molecules are broken down into smaller molecules by chemicals called **enzymes**. There are three main types of digestive enzymes. Each type of enzyme breaks down a different type of food.

Enzymes called **amylases** break down **starch** into **sugar**.
Enzymes called **lipases** break down **fats** into **fatty acids** and **glycerol**.
Enzymes called **proteases** break down **proteins** into **amino acids**.

Enzymes :

1 speed up digestion.
2 work best at the normal body temperature (37°C in man).
3 are affected by the acidity of alkalinity of the liquid around them.

Questions

1 Name the enzymes which digest starch.
2 What are lipases?
3 What are proteases?
4 At which temperature do enzymes work best in man?

Where food is digested

Food is digested in a long tube called the gut or **alimentary canal**.

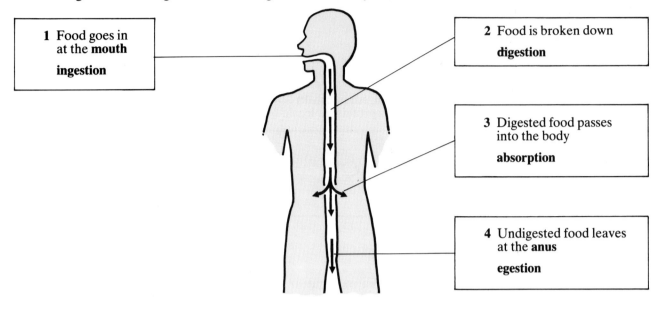

1 Food goes in at the **mouth**

ingestion

2 Food is broken down

digestion

3 Digested food passes into the body

absorption

4 Undigested food leaves at the **anus**

egestion

Digestion in the mouth

Food is chewed by the teeth to break it down into smaller pieces. As the food is chewed it is mixed with a digestive juice called **saliva.** The tongue helps to mix the food and saliva together. The small pieces of food are then made into a ball or **bolus** and swallowed.

Swallowing

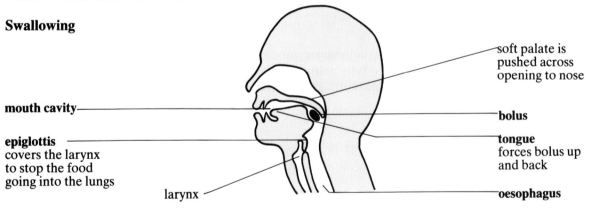

soft palate is pushed across opening to nose

mouth cavity

bolus

epiglottis
covers the larynx
to stop the food
going into the lungs

tongue
forces bolus up
and back

larynx

oesophagus

Saliva

Saliva is made in the **salivary glands** which are under the tongue and in the cheeks.

Saliva contains :

1 an amylase called **salivary amylase** which breaks down some starch into sugar.
2 a slimy substance called **mucin** which helps to soften the food.

The oesophagus

When the food is swallowed it goes into the oesophagus. The oesophagus has two sets of muscle fibres.

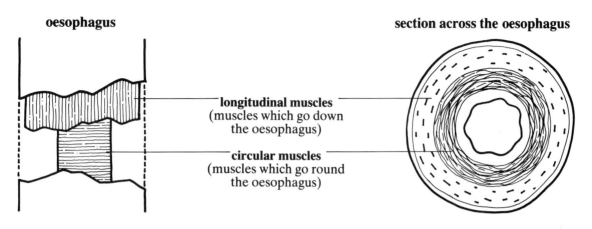

oesophagus

section across the oesophagus

longitudinal muscles
(muscles which go down
the oesophagus)

circular muscles
(muscles which go round
the oesophagus)

Peristalsis

When the circular muscles in the oesophagus contract the food is pushed towards the stomach. This is known as **peristalsis**. Food moves along the whole alimentary canal by peristalsis.

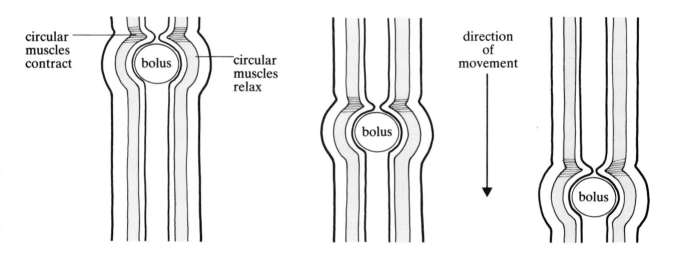

circular
muscles
contract

bolus

circular
muscles
relax

direction
of
movement

bolus

bolus

Questions

1 What is ingestion?
2 Why is food chewed?
3 (a) Which digestive juice is added to food in the mouth?
   (b) Name two substances found in this juice.
4 What is a bolus?
5 Why is the epiglottis important?
6 Name the two types of muscle in the wall of the oesophagus
7 What is peristalsis?

55

Digestion in the stomach

The oesophagus carries food to the stomach. The stomach is like a bag and the food may stay there for some time. The cells lining the stomach make **gastric juice** and **acid**. The food is mixed with gastric juice and acid when the muscles in the stomach wall contract.

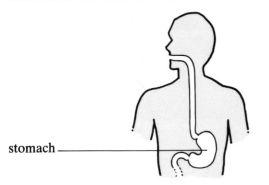

stomach

Gastric juice contains :

1 a protease called **pepsin** which starts the breakdown of protein.
2 a protease called **rennin** which turns milk into a solid.

The acid :

1 kills most bacteria
2 makes pepsin work
3 stops salivary amylase working.

Questions

1 Which digestive juice is made by the cells lining the stomach?
2 **(a)** Name the two enzymes which are added to the food in the stomach.
   (b) Explain what each enzyme does.
3 Is the liquid in the stomach acidic or alkaline? Give reasons for your answer.

Digestion in the small intestine

When food leaves the stomach it goes into the small intestine. The two parts of the small intestine are called the **duodenum** and the **ileum**.
   Bile and **pancreatic juice** are added to the food in the duodenum.
Intestinal juice is mixed with the food as it passes along the ileum.

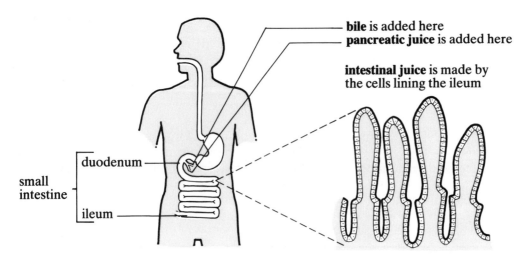

bile is added here
pancreatic juice is added here

intestinal juice is made by the cells lining the ileum

small intestine

duodenum

ileum

Bile

Bile is a green liquid which is made in the **liver**.

Bile contains **salts** which

1 make the food alkaline
2 make fat droplets smaller

Pancreatic juice

Pancreatic juice is made in the pancreas.

Pancreatic juice contains

1 an **amylase** (pancreatic amylase) which breaks down starch to sugar
2 a **protease** which breaks down protein into smaller molecules
3 a **lipase** which breaks down fats into **fatty acids** and **glycerol**

Intestinal juice

Intestinal juice contains

1 a **protease** which finishes breaking down proteins into **amino acids**
2 other **enzymes** which break down sugar to **glucose**

What happens to undigested food

Some of the food we eat cannot be digested. Vegetables and other foods which come from plants have cell walls made of **cellulose**. Although cellulose cannot be digested it is still an important part of the diet. The undigested cellulose or **roughage** is bulky and the muscles of the alimentary canal can push against this. Roughage and other undigested food in the ileum goes into the **large intestine** or **colon**.

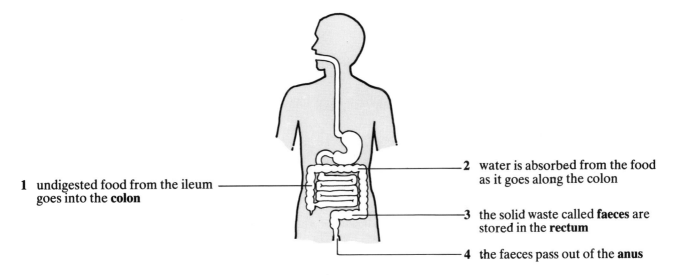

1 undigested food from the ileum goes into the **colon**

2 water is absorbed from the food as it goes along the colon

3 the solid waste called **faeces** are stored in the **rectum**

4 the faeces pass out of the **anus**

Questions

1 Name the two parts of the small intestine.
2 Where is pancreatic juice made?
3 How does pancreatic juice help with the digestion of food?
4 What is roughage?
5 Why is roughage an important part of the diet?
6 In which part of the large intestine is water absorbed?
7 What are faeces?
8 Where are faeces stored?

The digestive system of man

The alimentary canal is about five metres long. It is coiled round so that it can fit into the abdomen.

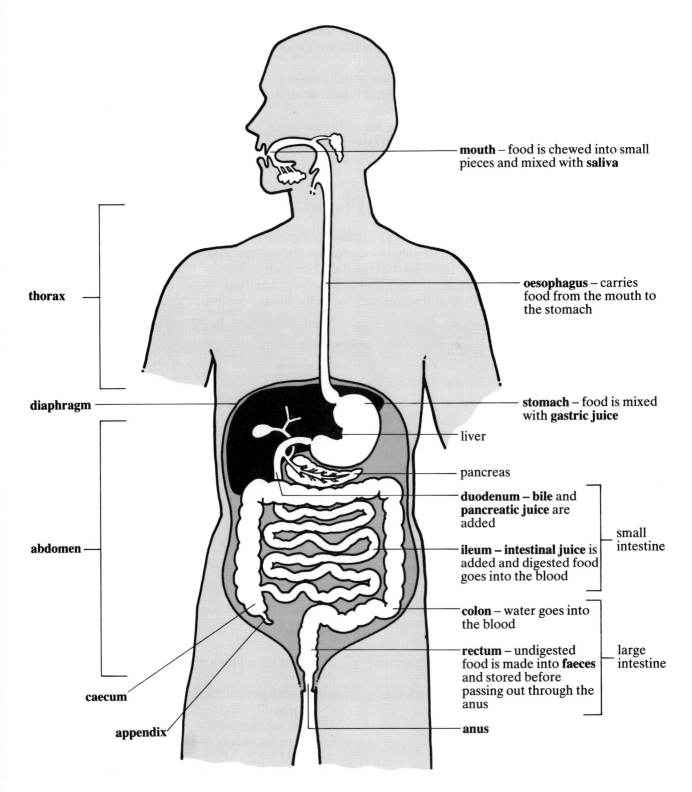

thorax

diaphragm

abdomen

caecum

appendix

mouth – food is chewed into small pieces and mixed with **saliva**

oesophagus – carries food from the mouth to the stomach

stomach – food is mixed with **gastric juice**

liver

pancreas

duodenum – bile and **pancreatic juice** are added

ileum – intestinal juice is added and digested food goes into the blood

small intestine

colon – water goes into the blood

rectum – undigested food is made into **faeces** and stored before passing out through the anus

large intestine

anus

The digestive system of a rabbit

Plants are difficult to digest because they have cell walls made of cellulose. Although a rabbit eats plants it cannot digest cellulose. The cellulose is digested by bacteria living in the rabbit's **caecum**. The bacteria break down the cellulose into substances which the rabbit can use.

In this diagram the parts of the digestive system have been spread out.

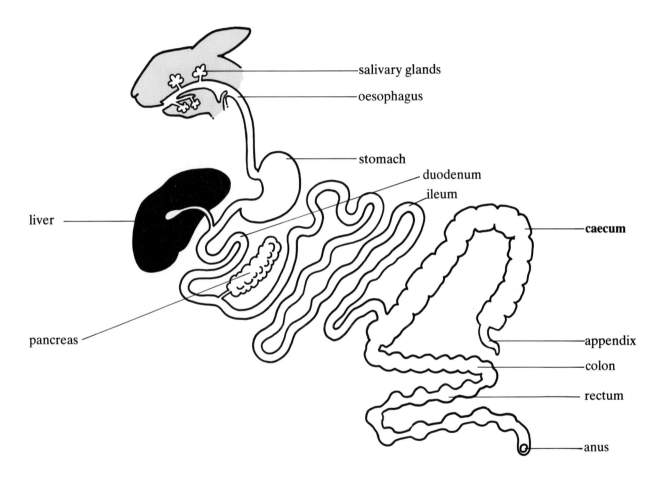

salivary glands
oesophagus
stomach
duodenum
ileum
caecum
liver
appendix
colon
rectum
pancreas
anus

Questions

1 What happens to food in the mouth?
2 Which part of the alimentary canal carries food from the mouth to the stomach?
3 What happens to the food in the ileum?
4 Copy the diagram of the digestive system of man and label all the parts which make digestive juices.
5 Why are plants difficult to digest?
6 Why does a rabbit need bacteria in its caecum?

Absorption

Some simple substances like glucose and alcohol are absorbed in the stomach. Most absorption takes place in the **ileum**. Glucose, amino acids, fatty acids and glycerol are absorbed in the ileum.

The ileum

The ileum has a large surface for absorbing food. It is very long and has small finger-like projections called **villi**.

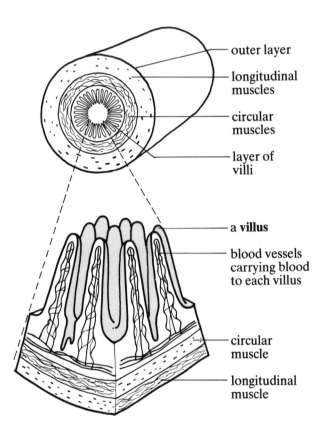

- outer layer
- longitudinal muscles
- circular muscles
- layer of villi
- a **villus**
- blood vessels carrying blood to each villus
- circular muscle
- longitudinal muscle

A villus

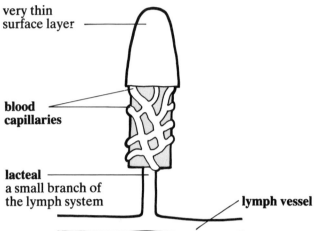

very thin surface layer

blood capillaries

lacteal a small branch of the lymph system

lymph vessel

Food is absorbed through the very thin surface layer.

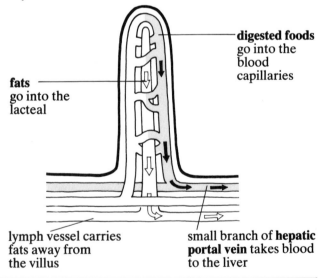

digested foods go into the blood capillaries

fats go into the lacteal

lymph vessel carries fats away from the villus

small branch of **hepatic portal vein** takes blood to the liver

---

Questions

1 Name two substances which can be absorbed in the stomach.
2 Name four substances which are absorbed in the ileum.
3 What are villi?

4 Which digested foods go into the lacteal?
5 Which digested foods go into the blood capillaries?
6 Name the vein which carries digested food to the liver.

What happens to digested foods

Digested foods are carried away from the ileum by the blood and lymph systems. The blood carries glucose and amino acids to the liver. The lymph carries digested fats to a vein at the base of the neck. The digested fats then pass into the blood system.

Glucose

Glucose may be – **1** used by the body for **energy**.
2 changed into **glycogen** and stored until it is needed.

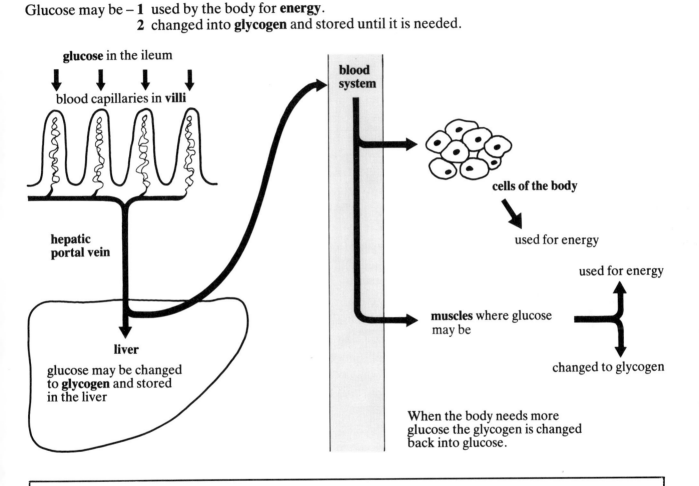

Questions

Copy and complete this paragraph by using the following words:

energy   blood   hepatic portal vein   muscles   glycogen   glucose

Glucose in the ileum goes into the............ capillaries in the villi. The blood is then carried by the............ to the liver. Glucose may be changed into............ and stored or carried by the blood to the cells where it is used for............ Some glucose may be changed into glycogen and stored in the............ When the body needs more energy the glycogen is changed back into............

Amino acids

Amino acids are used to make new proteins which are needed for growth and repair of the body.

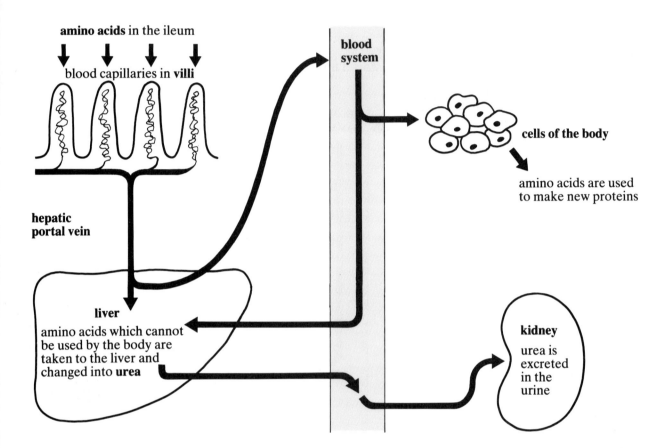

Questions

Copy and complete this paragraph by using the following words:

excreted   villi   blood   ileum   proteins   kidneys   urea   liver

Amino acids in the............ go into the blood capillaries in the............ The blood is then carried by the hepatic portal vein to the............ Amino acids are then carried by the............ to the cells. The amino acids are used to make new............ which can be used for growth and repair of the body. Any amino acids which cannot be used are taken to the liver and changed into............ The blood carries the urea to the............ so that it can be............ from the body.

Fatty acids and glycerol

Fats can be used for energy or stored in some parts of the body until they are needed. Some fatty acids and glycerol are absorbed by the blood in the villi. Other fats go into the lacteals.

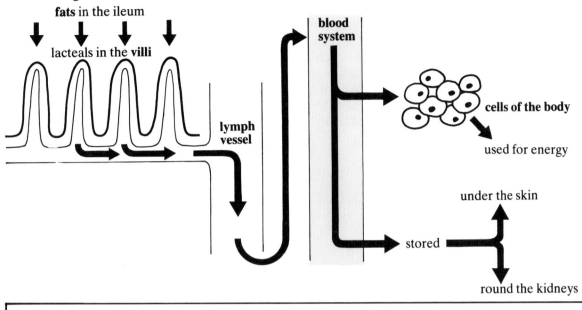

fats in the ileum

lacteals in the **villi**

blood system

lymph vessel

cells of the body

used for energy

under the skin

stored

round the kidneys

Questions

Copy and complete this paragraph by using the following words:

energy   kidneys   skin   cells   lacteals   stored

Some fats go into the............ which are small branches of the lymph system. The blood carries fats to the............ where they can be used for............ Some fat can be............ under the............ or round the heart and............ .

The liver

The liver helps to control the amount of glucose in the blood. If there is too much glucose in the blood some of it is changed into glycogen and stored in the liver. If there is not enough glucose in the blood some of the stored glycogen is changed back into glucose. The liver has many other functions.

The liver:

1 changes **glucose → glycogen** to store it.

2 changes **glycogen → glucose** when it is needed by the body.

3 changes **amino acids** which cannot be used by the body into **urea**.

4 stores **iron**.

5 makes **bile**.

6 removes some **poisons** from the blood.

heat is produced by these activities

Summary

The food we eat must be broken down before it can be used by the body. Digestion breaks down large food molecules into smaller ones which can be absorbed and used by the cells.

   Large pieces of food are broken down into smaller pieces by chewing. Digestive juices containing enzymes are added to the food as it passes along the alimentary canal. The enzymes break down starch to glucose; proteins to amino acids; fats to fatty acids and glycerol.

   Glucose is used for energy. Amino acids are used to make new proteins. Fats are either used for energy or stored until they are needed.

   Water is absorbed from the undigested food before it passes out through the anus.

Key words

absorption   Taking digested food into the blood

digestion   Breaking down food so that it can be used by the body

enzyme   Enzymes break down large food molecules into smaller molecules. The three types of digestive enzymes are
amylases which break down starch to glucose.
lipases which break down fats to fatty acids and glycerol
proteases which break down proteins to amino acids

Questions

1 Why is digestion necessary?

2 (a) What are enzymes?
   (b) Copy and complete this table. Put + to show which enzymes are found in each digestive juice. The first one is done for you.

Digestive juice	Amylase	Lipase	Protease
saliva	+	–	–
gastric juice			
pancreatic juice			
intestinal juice			

3 Why is roughage an important part of the diet?

4 Describe what happens to the food in the mouth.

5 How is food moved from one region of the alimentary canal to another?

6 Explain the following statement:

A boy standing on his head can drink water and this will go to his stomach, once it is swallowed.

7 What gives the ileum a large surface area?

8 Why does the ileum need a large surface area?

9 Explain why the liver is an important part of the body.

10 How is glucose used by the body?

11 What happens to amino acids not used for body building or repair?

12 How does the body use fats?

13 Explain how a rabbit can digest cellulose.

14 Why do you think some mammals produce rennin when they are young but not when they are older?

15 Describe the digestion of a piece of bread (starch). What happens after the food has been absorbed into the blood? What is it eventually used for?

16 An experiment was set up to show what happens to starch in the mouth.

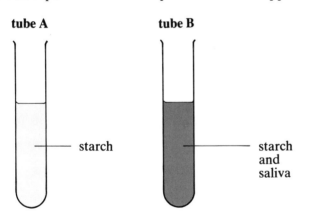

Both tubes were kept in a water bath at 37°C for 30 minutes.

After 30 minutes both solutions were tested for starch and glucose.

Results after 30 minutes

	tube A	tube B
starch	✓	✗
glucose	✗	✓

(a) Why were the tubes kept in a water bath at 37°C?

(b) Explain why there was glucose in tube B after 30 minutes.

(c) Why did the starch in tube A not change into glucose?

(d) Explain the following statement:

A girl chewed some white bread for a short time and said there was a sugary taste in her mouth.

(e) If starch and glucose are eaten, only the starch will have to be digested. Explain why this is so.

9: Different ways of life

Some living things are called **parasites**. A parasite lives on or in another living animal or plant which is called the **host**. The host is usually harmed in some way when the parasite feeds.

There are two types of animal parasites:

1 **external parasites**
2 **internal parasites**

External parasites

External parasites live and feed outside the host. Ticks, fleas and lice suck blood from their hosts. They may spread disease when they feed.

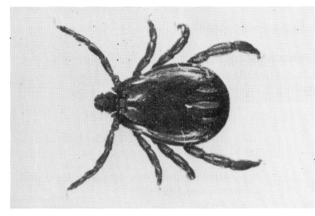

A **tick** lives on cattle. Ticks carry diseases from one cow to another.

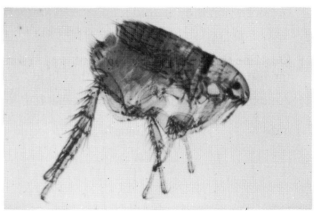

A **flea** has strong legs for jumping from one host to another.

A **louse** lives on human hair. It uses claws to cling on to its host.

Internal parasites

Internal parasites live and feed inside the host. Tapeworms and liver flukes
are internal parasites. Both animals are suited to a parasitic way of life. They:

1 are **protected** by a tough **cuticle** covering the whole body.
2 can **attach** themselves **to the host**.
3 have both **male and female sex organs**.
4 produce **many eggs** because only a few will find another host.

A **tapeworm** lives inside the
intestines of man.

A **liver fluke** lives in the liver of
sheep.

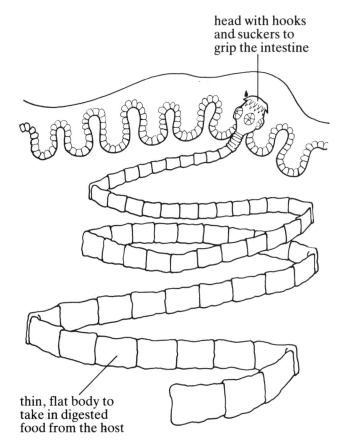

head with hooks
and suckers to
grip the intestine

thin, flat body to
take in digested
food from the host

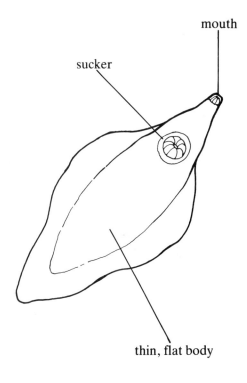

mouth

sucker

thin, flat body

Questions

1 What is a parasite?
2 What is meant by a 'host'?
3 Name three external parasites.
4 Name two internal parasites.
5 How is a tapeworm adapted to a parasitic way of life?

The liver fluke

The liver fluke needs two different hosts to complete its life cycle.

1 The adult liver fluke lives inside a **sheep**. The sheep is called the **primary host**.
2 The larva of the liver fluke lives inside a **water snail**. The water snail is called the **secondary host**.

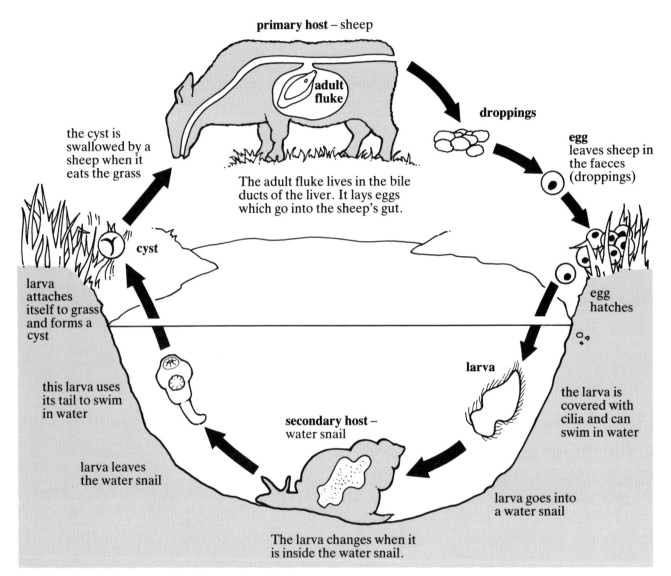

primary host – sheep

adult fluke

droppings

egg
leaves sheep in the faeces (droppings)

the cyst is swallowed by a sheep when it eats the grass

The adult fluke lives in the bile ducts of the liver. It lays eggs which go into the sheep's gut.

cyst

egg hatches

larva attaches itself to grass and forms a cyst

larva

this larva uses its tail to swim in water

the larva is covered with cilia and can swim in water

larva leaves the water snail

secondary host – water snail

larva goes into a water snail

The larva changes when it is inside the water snail.

How to stop sheep being infected by liver flukes

Liver flukes cause 'liver rot' in sheep. A sheep may die from 'liver rot'. To stop their sheep being infected by liver flukes farmers may:

1 kill the water snails.
2 keep the sheep away from wet fields.
3 drain ponds.
4 keep ducks and geese which eat the water snails.

Plant parasites

Some plants are parasites. **Dodder** is a parasitic plant. Dodder gets the food, water and minerals it needs by living on another plant.

Dodder

Host plant – heather

Parasite – dodder
Dodder looks like thin, pink cotton coiled round the stem of its host.

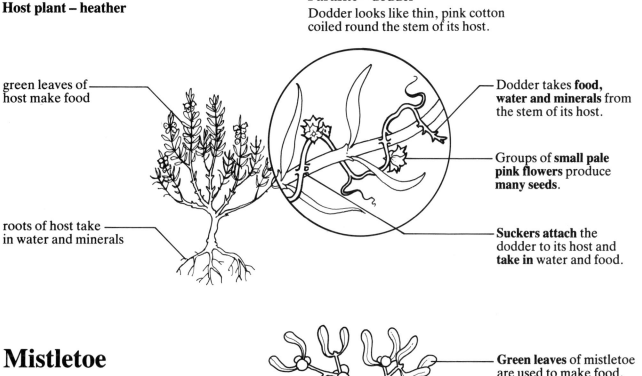

green leaves of host make food

Dodder takes **food, water and minerals** from the stem of its host.

Groups of **small pale pink flowers** produce **many seeds**.

roots of host take in water and minerals

Suckers attach the dodder to its host and **take in** water and food.

Mistletoe

The mistletoe is a **semi-parasite**. It takes water and minerals from the host but it can make its own food.

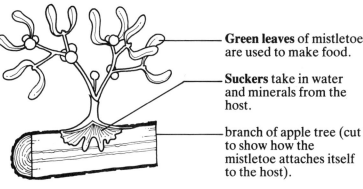

Green leaves of mistletoe are used to make food.

Suckers take in water and minerals from the host.

branch of apple tree (cut to show how the mistletoe attaches itself to the host).

Questions

1 Do liver flukes harm a sheep? Give reasons for your answer.
2 What happens to the larva of a liver fluke when it is inside a water snail?
3 Name the primary host of a liver fluke.
4 What can a farmer do to stop his sheep being infected by the liver fluke?
5 Why is dodder a parasite?
6 Name the dodder's host.
7 Why must the dodder produce a large number of seeds?
8 Look carefully at the diagram of the dodder. It has a very thin stem. How is the dodder supported?

Saprophytes

Some fungi and bacteria feed on the dead remains of plants and animals. These fungi are called **saprophytes**. **Mucor** is a saprophytic fungus.

Mucor (pin mould)

Mucor is a small fungus which grows on bread and many other things. This diagram shows Mucor growing on a piece of bread. It has been magnified to show the different parts of Mucor.

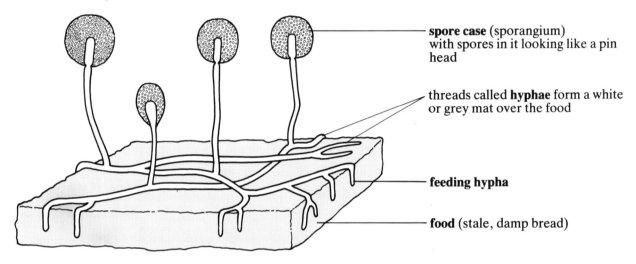

spore case (sporangium) with spores in it looking like a pin head

threads called **hyphae** form a white or grey mat over the food

feeding hypha

food (stale, damp bread)

How mucor feeds

Mucor feeds on bread, fruit and manure. This diagram shows part of a feeding hypha in some food.

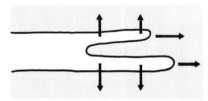 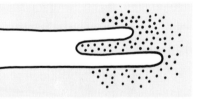 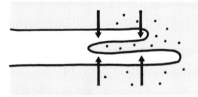

1 hypha secretes **enzymes** on to the food

2 enzymes **digest** the food

3 **digested food** goes **into the hypha**

Why saprophytes are important

1 The soil contains many saprophytic fungi and bacteria. When saprophytes feed on the dead remains of plants and animals important chemicals are released. Saprophytes speed up decay and keep these chemicals circulating in nature (see chapter 21).
2 Saprophytic bacteria help to change sewage into harmless substances.
3 The antibiotic penicillin comes from a saprophytic mould.

Insect-eating plants

Some plants make their own food, but get an extra supply of nitrogen by trapping and 'eating' insects. These plants are called **insectivorous plants**. They can live without eating insects, but they grow larger and produce more flowers if they get this extra food.

The sundew

The sundew grows in bogs and other wet places where there is very little nitrogen in the soil.

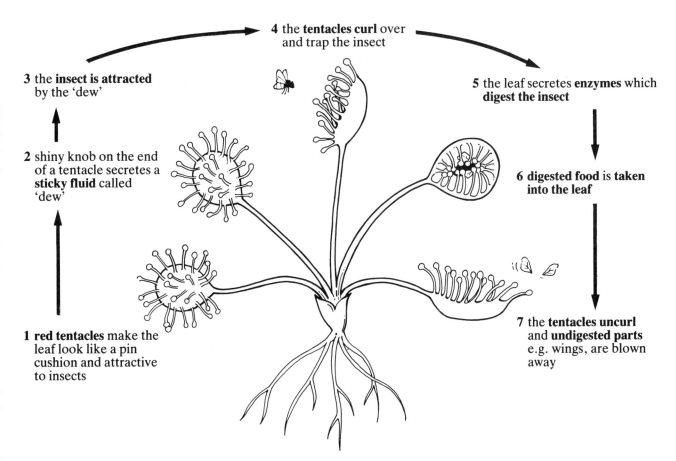

4 the **tentacles curl** over and trap the insect

3 the **insect is attracted** by the 'dew'

2 shiny knob on the end of a tentacle secretes a **sticky fluid** called 'dew'

1 red tentacles make the leaf look like a pin cushion and attractive to insects

5 the leaf secretes **enzymes** which **digest the insect**

6 digested food is **taken into the leaf**

7 the **tentacles uncurl** and **undigested parts** e.g. wings, are blown away

---

Questions

1 What is a saprophyte?
2 Name one saprophyte.
3 Explain how a saprophyte feeds.
4 Why are saprophytes important?
5 Name an insectivorous plant.
6 In your own words, explain how one insectivorous plant you have studied traps and 'eats' insects.

Symbiosis

Sometimes two completely different plants or animals live together. If both benefit from living together it is called **symbiosis**.

Both the shark and the pilot fish benefit from living together. Pilot fish often swim beside a shark. The pilot fish are protected by the shark because they swim into its mouth when there is danger. The shark has its mouth cleaned by the pilot fish when they eat the scraps of food left in the mouth.

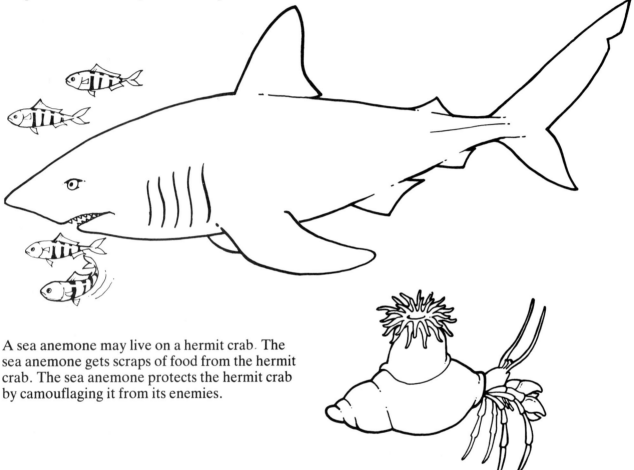

A sea anemone may live on a hermit crab. The sea anemone gets scraps of food from the hermit crab. The sea anemone protects the hermit crab by camouflaging it from its enemies.

Summary

All living things need food to stay alive. Parasites get their food from other living plants or animals. A parasite harms its host in some way. Saprophytes feed on dead and decaying matter. A saprophyte secretes enzymes to break down the food before it is taken into the body.

If two completely different plants or animals benefit from living together it is called symbiosis.

Key words

host An animal or plant which has a parasite feeding on it

insectivorous plant A plant which traps and digests insects to get extra nitrogen e.g. sundew

parasite An animal or plant which feeds on another living thing e.g. liver fluke, tapeworm

saprophytes Fungi or bacteria which feed on dead and decaying plant and animal remains e.g. Mucor. Saprophytes digest their food before taking it into the body

symbiosis When two different plants or animals benefit from living together

Questions

1 What is symbiosis?

2 Give one example of symbiosis.

3 Name an internal parasite you have studied.
   (a) Explain how it gets its food.
   (b) Describe how it is adapted to its parasitic way of life.
   (c) How is the host affected by the parasite?
   (d) Describe its life cycle (you may use diagrams to help explain what happens).
   (e) What could be done to stop the parasite reaching its primary host?

4 How is the dodder adapted to its parasitic way of life?

5 How do saprophytes help to keep certain chemical elements circulating in nature?

6 A boy put a sandwich inside a polythene bag and left it in his desk. He forgot to eat the sandwich and a few days later he noticed white and grey patches on the bread.
   (a) Draw what you would expect to see if one of the 'patches' was put under a strong magnifying glass.
   (b) Label as many parts as you can.
   (c) Name the organism forming the 'patches'.
   (d) Explain how the organism feeds.

7 Why do parasites produce a large number of eggs or seeds?

8 What is the main difference between mistletoe and dodder?

9 In warmer countries a farmer sprays his cattle every few weeks because a cow may carry several hundred ticks.
   (a) How does the tick feed?
   (b) Is the cow harmed in any way? Explain your answer.

10 You may have suffered from a skin infection called athlete's foot. Athlete's foot is caused by a fungus. What method of nutrition would this fungus use? Give reasons for your answer.

11 What is the basic difference between a saprophyte and a green plant?

12 Water cress is grown in slow flowing streams. Many snails are found in the water. Man eats water cress with salads. Explain why it is important to wash water cress before eating it.

10: Respiration and gas exchange

All living things need **energy**. Every process that goes on inside the body uses energy. Plants and animals get their energy from food. The energy stored in food is released by **respiration**.

Respiration

Respiration takes place in the cells of plants and animals. Respiration releases energy from sugar.

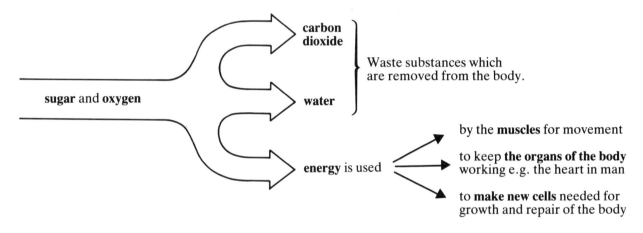

sugar and **oxygen** →

carbon dioxide

water

} Waste substances which are removed from the body.

energy is used →

by the **muscles** for movement

to keep **the organs of the body** working e.g. the heart in man

to **make new cells** needed for growth and repair of the body

Aerobic and anaerobic respiration

Aerobic respiration

When **oxygen** is used to release energy from sugar it is called **aerobic respiration**.

sugar + oxygen → carbon dioxide + water + energy

Anaerobic respiration

When muscles are working hard they use up a lot of oxygen. If oxygen does not get to the muscles fast enough for aerobic respiration they change to **anaerobic respiration**.

sugar → carbon dioxide + lactic acid + some energy

Lactic acid builds up in the muscles and stops them working. The person breathes faster to take in more oxygen so that the lactic acid can be broken down to carbon dioxide and water.

Fermentation

Fermentation is a kind of anaerobic respiration. In fermentation alcohol is produced instead of lactic acid.

Yeast ferments sugar into **alcohol** and **carbon dioxide**. Man uses yeast to make alcoholic drinks and bread.

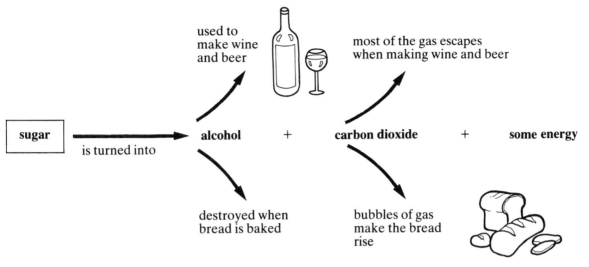

used to make wine and beer

most of the gas escapes when making wine and beer

sugar → is turned into → alcohol + carbon dioxide + some energy

destroyed when bread is baked

bubbles of gas make the bread rise

Questions

1 Where do living organisms get their energy from?
2 Which gas combines with sugar to release energy?
3 Where does respiration take place?
4 Explain why living organisms need energy.
5 Name two waste substances made by aerobic respiration.
6 What is anaerobic respiration?

Gas exchange

When plants and animals respire aerobically they exchange gases with the air or water around them. They **take in oxygen** and **give out carbon dioxide**.

Gas exchange in Amoeba

Amoeba is an animal which has only one cell. Gases are exchanged through the whole surface of the cell. Every part of the cell is near the water. Oxygen which is dissolved in the water can get to every part of the cell by diffusion (see page 17). Carbon dioxide made in the cell can get to the surface by diffusion.

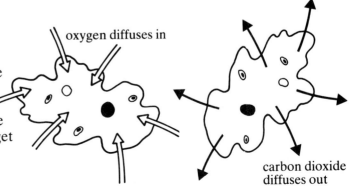

oxygen diffuses in

carbon dioxide diffuses out

Gas exchange in insects

Insects are made up of many cells. Some of the cells are a long way from the air. Oxygen cannot reach all the cells by diffusion. Gases are carried to and from the cells by a network of thin tubes. This network of tubes is called a **tracheal system**. Air is taken into the system, and squeezed out of the system, by movements of the abdomen.

The tracheal system of a locust

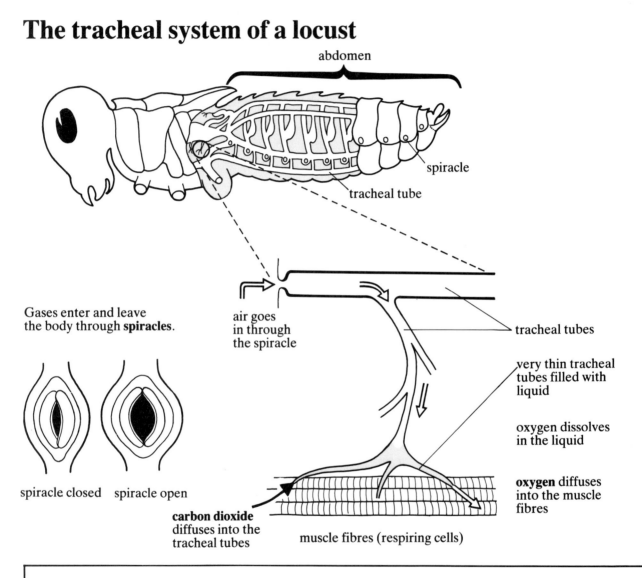

abdomen

spiracle

tracheal tube

Gases enter and leave the body through **spiracles**.

air goes in through the spiracle

tracheal tubes

very thin tracheal tubes filled with liquid

oxygen dissolves in the liquid

oxygen diffuses into the muscle fibres

spiracle closed   spiracle open

carbon dioxide diffuses into the tracheal tubes

muscle fibres (respiring cells)

Questions

1 Where does gas exchange take place in an Amoeba?
2 How does oxygen reach every part of an Amoeba?
3 Where do gases enter and leave the body of a locust?
4 What is a tracheal system?
5 Name an organism which has a tracheal system.

Gas exchange in an earthworm

The earthworm uses its skin as a **respiratory surface**. The skin takes in enough oxygen for all the cells in the body. Oxygen is carried to the cells by the blood.

Gases are exchanged through the whole surface of the skin.

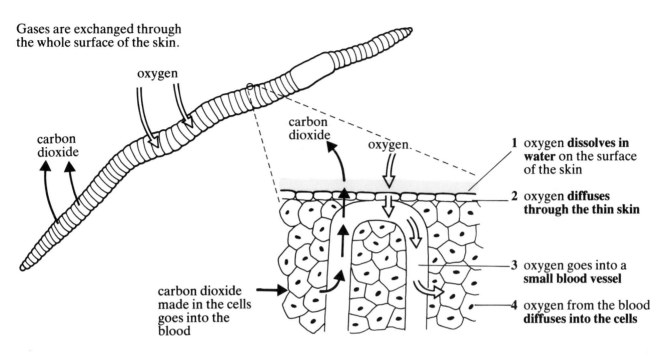

oxygen

carbon dioxide

carbon dioxide

oxygen.

carbon dioxide made in the cells goes into the blood

1 oxygen **dissolves in water** on the surface of the skin

2 oxygen **diffuses through the thin skin**

3 oxygen goes into a **small blood vessel**

4 oxygen from the blood **diffuses into the cells**

Respiratory surfaces

Large animals have respiratory surfaces which are:

1 **well supplied with air or water**.
2 **moist** because gases must be dissolved in water before they can go into or come out of a cell.
3 **large** so that enough oxygen can be taken into the body.
4 **well supplied with blood**. The blood carries oxygen from the respiratory surface to the cells. Blood also carries carbon dioxide from the cells to the respiratory surface.

Most large animals have a respiratory surface inside the body. The respiratory surface is often part of the **respiratory organs**. These are special parts of the body used for gas exchange.

Questions

1 Where does gas exchange take place in an earthworm?
2 Explain how oxygen gets to the cells of an earthworm.
3 Why are respiratory surfaces moist?
4 Why are respiratory surfaces in large animals well supplied with blood?

The respiratory organs of a fish

The respiratory organs of a fish are called **gills**. Gases are exchanged as water passes over the gills.

1 water with oxygen dissolved in it goes in through the mouth

2 gases are exchanged as the water passes over the gills

3 water with a lot of carbon dioxide in it comes out of the operculum

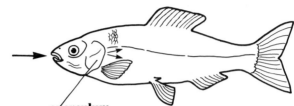

operculum
(a bony plate covering the gills)

Gills

In this diagram the operculum has been cut open to show the gills.

gills

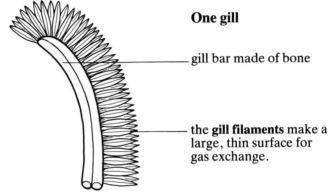

One gill

gill bar made of bone

the **gill filaments** make a large, thin surface for gas exchange.

The **gill filaments** are well supplied with blood. Oxygen diffuses into the blood as the water passes over the gill filaments. At the same time carbon dioxide diffuses into the water.

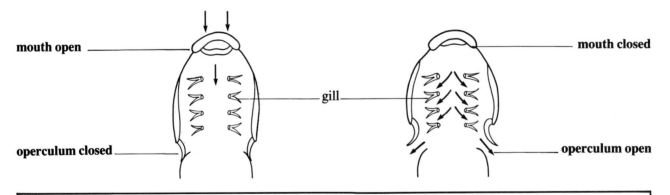

How water goes into the mouth

mouth open

operculum closed

gill

How water leaves the mouth

mouth closed

operculum open

Questions

1 What are the respiratory organs of a fish called?

2 Explain why the respiratory organs of a fish are well supplied with blood.

3 What is an operculum?

4 Explain what happens to the mouth and operculum when **(a)** water goes into the mouth **(b)** water leaves the mouth.

The respiratory system of man

The **lungs** are the respiratory organs of man. The lungs are in the chest or **thorax**.

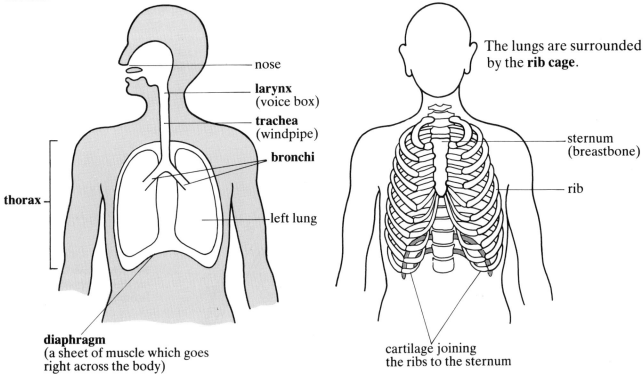

The lungs are surrounded by the **rib cage**.

nose

larynx (voice box)

trachea (windpipe)

bronchi

left lung

thorax

diaphragm (a sheet of muscle which goes right across the body)

sternum (breastbone)

rib

cartilage joining the ribs to the sternum

The nose

The air is moistened and warmed as it goes through the nose. The cells which line the nose have **cilia** on them. Cilia look like tiny hairs. When the cilia move they push **mucus**, dust and dirt towards the back of the mouth. This helps to clean the air before it goes into the lungs.

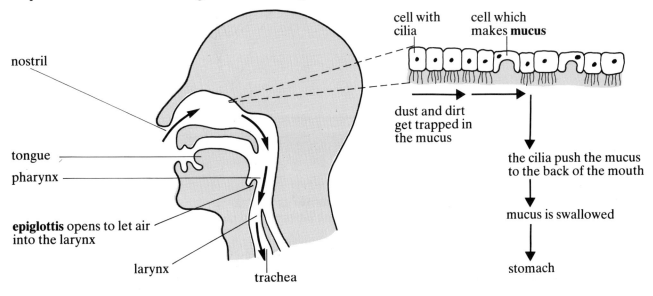

cell with cilia

cell which makes **mucus**

nostril

tongue

pharynx

epiglottis opens to let air into the larynx

larynx

trachea

dust and dirt get trapped in the mucus

the cilia push the mucus to the back of the mouth

mucus is swallowed

stomach

Section through the thorax

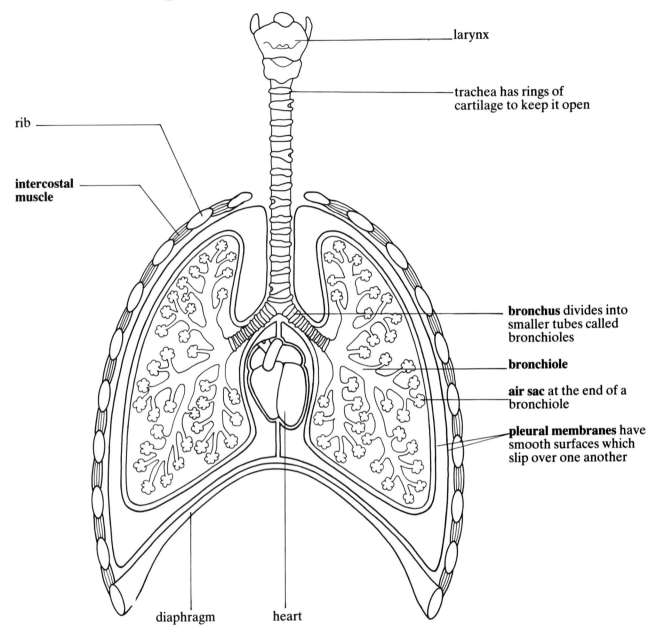

larynx

trachea has rings of cartilage to keep it open

rib

intercostal muscle

bronchus divides into smaller tubes called bronchioles

bronchiole

air sac at the end of a bronchiole

pleural membranes have smooth surfaces which slip over one another

diaphragm

heart

Questions

1 What are the respiratory organs of man called?
2 What is the diaphragm?
3 What happens to the air as it goes through the nose?
4 Explain what happens to the dust and dirt in the air breathed in through the nose.
5 Why are there rings of cartilage round the trachea?

Air sacs

Each air sac is made up of many **alveoli**. The alveoli make a large moist surface for gas exchange. The alveoli are well supplied with blood.

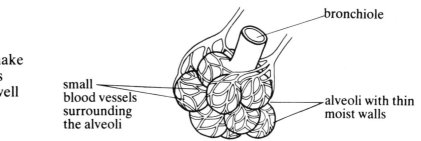

bronchiole

small blood vessels surrounding the alveoli

alveoli with thin moist walls

Gas exchange in the alveoli

Oxygen from the air in the alveoli diffuses into the blood. At the same time, carbon dioxide from the blood diffuses into the alveoli.

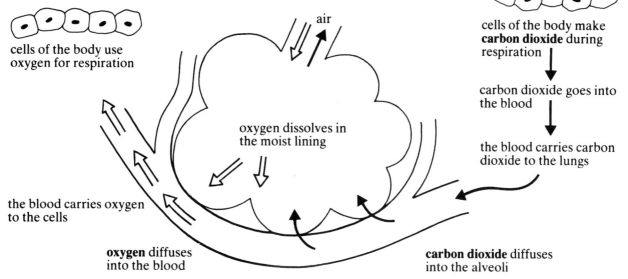

cells of the body use oxygen for respiration

air

oxygen dissolves in the moist lining

the blood carries oxygen to the cells

oxygen diffuses into the blood

cells of the body make **carbon dioxide** during respiration

carbon dioxide goes into the blood

the blood carries carbon dioxide to the lungs

carbon dioxide diffuses into the alveoli

How air changes when it is in the lungs		
gas	air going into the lungs (inspired air)	air coming out of the lungs (expired air)
oxygen	21%	16%
carbon dioxide	0·04%	4%
nitrogen	79%	79%
water vapour	a little	a lot

Questions

1 In which part of a lung does gas exchange take place?
2 How does the carbon dioxide get from the cells to the lungs?
3 What happens to the oxygen in the alveoli?
4 Give two differences between the air going into the lungs and the air coming out of the lungs.

Breathing

Air goes in and out of the lungs when the **intercostal muscles** (muscles in between the ribs) and **diaphragm** contract and relax.

Breathing in (inspiration)

Breathing out (expiration)

air rushes in

air is pushed out

intercostal muscles contract

trachea

rib cage moves up and out

rib

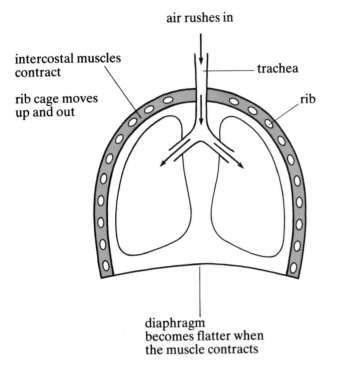

intercostal muscles relax

rib cage moves down and in

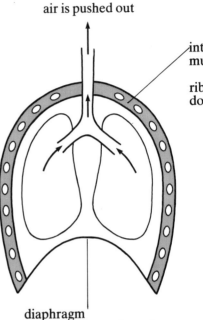

diaphragm becomes flatter when the muscle contracts

diaphragm becomes dome-shaped when the muscle relaxes

How breathing is controlled

Breathing is an unconscious action (not controlled by the mind). The normal breathing rate is 16 per minute. The breathing rate is controlled by part of the brain. When there is a lot of carbon dioxide in the blood the brain sends a message to the diaphragm and intercostal muscles to make them contract faster. This makes the person breathe faster. By breathing faster more carbon dioxide is removed from the lungs. At the same time more oxygen is taken into the lungs.

Questions

1 Where are the intercostal muscles found?
2 Explain what happens when you (a) breathe in and (b) breathe out.
3 What is the normal breathing rate?
4 Which part of the body controls the breathing rate?

Gas exchange in plants

Plants do not have special respiratory surfaces or respiratory organs. Oxygen from the air can diffuse to every cell. Carbon dioxide can diffuse from every cell to the air. Gases enter and leave a plant through pores in the leaves. The pores are called **stomata**. Each one of the pores is called a **stoma**.

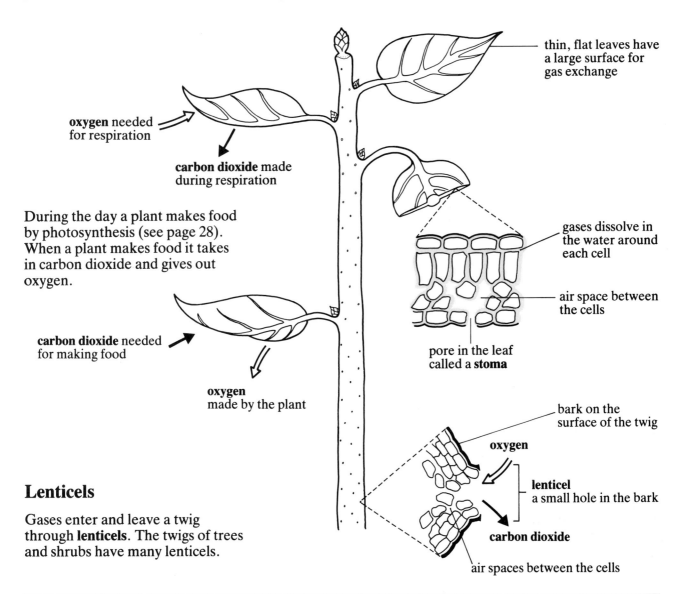

thin, flat leaves have a large surface for gas exchange

oxygen needed for respiration

carbon dioxide made during respiration

During the day a plant makes food by photosynthesis (see page 28). When a plant makes food it takes in carbon dioxide and gives out oxygen.

gases dissolve in the water around each cell

air space between the cells

carbon dioxide needed for making food

oxygen made by the plant

pore in the leaf called a **stoma**

bark on the surface of the twig

oxygen

Lenticels

Gases enter and leave a twig through **lenticels**. The twigs of trees and shrubs have many lenticels.

lenticel a small hole in the bark

carbon dioxide

air spaces between the cells

Questions

1 How does the shape of a leaf help gas exchange?
2 What are stomata?
3 Which gas does a plant use during respiration?
4 Which gas does a plant use to make its food?
5 What is a lenticel?

Summary

Energy is released when oxygen reacts with sugar in the cells of plants and animals. The energy released from food can be used for movement, growth and repair of the body.

Plants and animals must take in oxygen from the air or water around them. At the same time they must remove carbon dioxide. Gas exchange takes place through the whole cell surface of small animals like Amoeba. In most larger animals gas exchange takes place through special respiratory surfaces which may be part of respiratory organs such as gills in fish or lungs in mammals.

Respiratory surfaces are moist because gases must be dissolved in water before they can enter or leave living cells. Respiratory surfaces are well supplied with blood, which carries the gases to and from the cells.

Animals with respiratory surfaces inside the body must be able to take air into the body. Breathing takes air into the body. Breathing also speeds up the rate of gas exchange by removing carbon dioxide from the respiratory surface and supplying more oxygen as fast as it is taken in.

Plants do not have special respiratory organs. Gases enter and leave through stomata on the leaves and lenticels on woody stems. Gases move to and from all parts of a plant by diffusion.

Key words

aerobic respiration	Respiration which uses oxygen
alveoli	Part of the lung used for gas exchange
anaerobic respiration	Respiration which does not use oxygen
diaphragm	A sheet of muscle which separates the thorax and abdomen of a mammal
fermentation	A type of anaerobic respiration used by yeast and some bacteria
lenticel	A small hole in the bark of a twig. Gases enter and leave a woody twig through lenticels.
operculum	The bony plate covering a fish gill
respiratory surface	The part of an animal which is used for gas exchange. It may be part of a respiratory organ such as a gill or lung.
spiracle	An opening on an insect's body. Gases enter and leave through the spiracles.
stoma	A small pore in a leaf. One pore is a stoma. Many pores are stomata. Gases enter and leave through the stomata.
tracheal system	A network of tubes which carries air in and out of an insect's body

Questions

1 Which type of respiration breaks down sugar to (a) carbon dioxide and water and (b) alcohol and carbon dioxide?

2 An experiment was set up to show that carbon dioxide is produced during respiration.

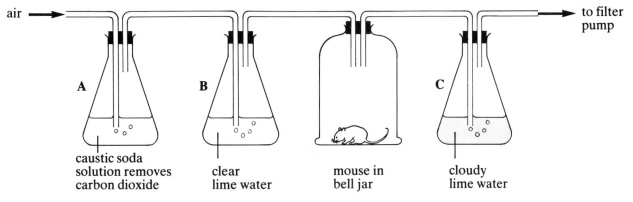

air →

to filter pump

A caustic soda solution removes carbon dioxide

B clear lime water

mouse in bell jar

C cloudy lime water

Lime water turns cloudy when there is carbon dioxide in the air.

(a) What happened to the air when it went through flask A?
(b) Why did the lime water in flask B stay clear?
(c) Why did the lime water in flask C go cloudy?

3 Three boys carried out an experiment to see if the breathing rate is affected by exercise. John, David and Robert recorded their breathing rates at rest. Then they ran round the school playground. When they returned to the classroom they recorded their breathing rates.

Table of breathing rates								
pupil	breathing rate before exercise	minutes after exercise						
		1	2	3	4	5	6	
John	15	43	36	31	22	17	15	per minute
Robert	14	41	35	30	20	15	14	per minute
David	13	31	21	14	13	13	13	per minute

(a) On squared paper, draw graphs to show the breathing rates of the three boys.
(b) What effect does running have on the breathing rate?
(c) Why does the breathing rate change during exercise?
(d) How long does it take for each boy to recover from the effects of exercise?

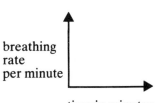

breathing rate per minute

time in minutes

4 This apparatus was set up to show that energy is released by respiration.

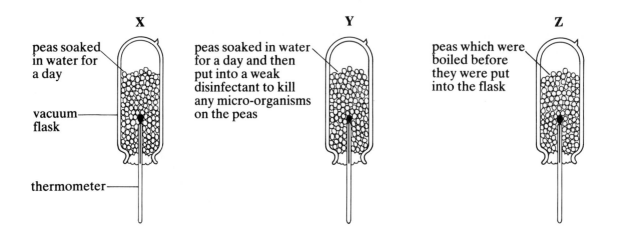

The flasks were left for three days. The following temperature changes were recorded (an increase in temperature shows that energy is being released).

	temperatures		
time	flask X	flask Y	flask Z
start of experiment	17°C	17°C	17°C
12 hours later	28°C	18°C	17°C
24 hours later	39°C	20°C	17°C
36 hours later	45°C	23°C	17°C
48 hours later	51°C	27°C	17°C
60 hours later	54°C	31°C	17°C
72 hours later	55°C	40°C	17°C

(a) Why did the temperature in flask X increase?
(b) Why did the temperature increase more in flask X than in Flask Y?
(c) Explain why the temperature in flask Z did not change.
(d) On squared paper draw a graph to show these results.

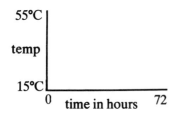

From your graph answer questions **(e)** and **(f)**.

(e) What was the temperature of flask X after 30 hours and after 42 hours?
(f) What was the temperature of flask Y after 18 hours and after 54 hours?

11: Transport in plants

Every plant cell needs water, minerals and sugar. Water and minerals are taken in by the roots and then carried to all the other parts of the plant. Sugar is made in the leaves and then carried to all the other parts of the plant. Water, minerals and sugar are carried from one part of the plant to another in very thin tubes. The tubes are grouped together in **vascular bundles**.

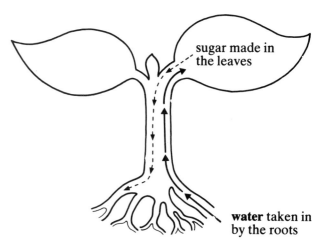

sugar made in the leaves

water taken in by the roots

The transport system in a stem

The vascular bundles are in a ring near the outside of a stem.

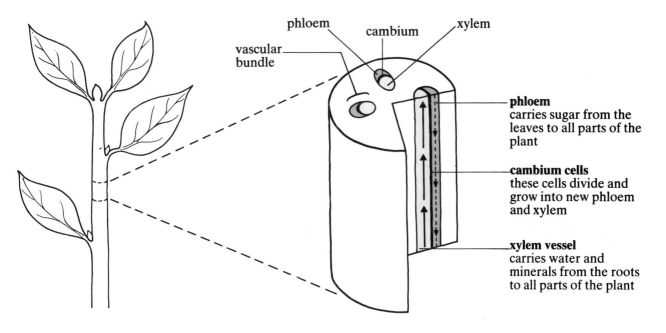

phloem

cambium

xylem

vascular bundle

phloem
carries sugar from the leaves to all parts of the plant

cambium cells
these cells divide and grow into new phloem and xylem

xylem vessel
carries water and minerals from the roots to all parts of the plant

Questions

1 Why does a plant need a transport system?
2 Name the three parts of a vascular bundle.
3 Explain what each part of a vascular bundle does.

The transport system in a root

The vascular bundles are grouped together in the middle of a root.

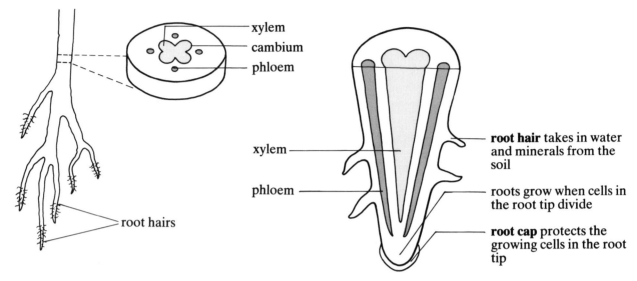

How water is taken in by the root hairs

Plant roots take in water and minerals from the soil. The minerals are dissolved in the water taken in by the root hairs.

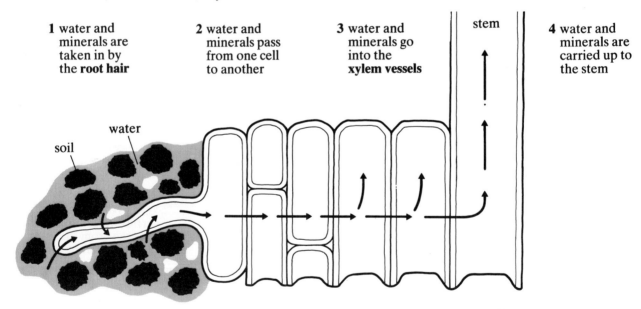

1 water and minerals are taken in by the **root hair**

2 water and minerals pass from one cell to another

3 water and minerals go into the **xylem vessels**

4 water and minerals are carried up to the stem

Questions

1 Which part of a root takes in water and minerals from the soil?
2 What does the root cap do?

Transpiration

A plant loses water through its leaves. The water is lost through small pores called **stomata**. This loss of water is called **transpiration**.

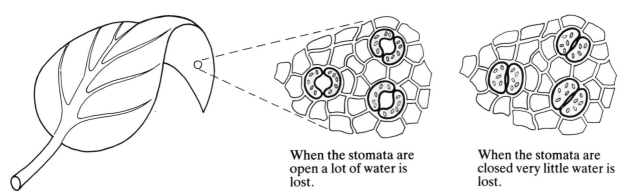

When the stomata are open a lot of water is lost.

When the stomata are closed very little water is lost.

The transpiration stream

As water is lost through the stomata more water is taken in by the roots. There is a continuous flow of water from the roots to the leaves. This is called the **transpiration stream**. The transpiration stream moves faster on warm, sunny days than it does on cold, cloudy days. If water is lost from the leaves faster than it is taken in by the roots the cells become limp and the plant **wilts**.

The transpiration stream is important because it carries:

1 **water** to the leaves so that the plant can make its food (photosynthesis).
2 **minerals** to the leaves for making proteins.
3 **water** to the cells which help to support the plant (see page 106).

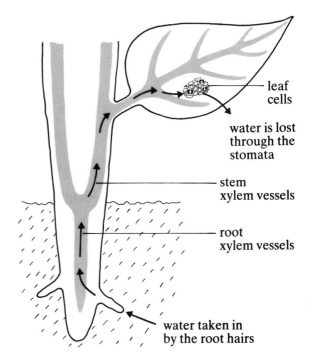

leaf cells

water is lost through the stomata

stem xylem vessels

root xylem vessels

water taken in by the root hairs

Questions

1 What is transpiration?
2 (a) What are stomata?
  (b) Where are stomata found?
3 Explain why the transpiration stream is important.
4 What happens if water is lost from the leaves faster than it is taken in by the roots?

Movement of water in a plant

Water moves from one cell to another by **osmosis** (see page 11). As water is lost from a cell the solution inside becomes stronger. This draws water out of the next cell.

A plant is made up of many cells. If you imagine that the three cells in the next diagram are the cells from the roots to the leaves you should be able to understand how the water lost through the leaves is replaced by water taken in by the roots.

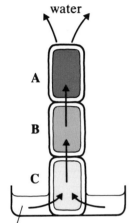

weak solution of minerals

As water is lost the solution inside cell A becomes stronger than that inside cell B. This draws water out of cell B.

As water is drawn out of cell B the solution inside cell B becomes stronger than that inside cell C. This draws water out of cell C.

As water is drawn out, the solution inside cell C becomes stronger than that in the dish. This draws water out of the dish.

Summary

Water and minerals taken in by the roots are carried to every part of a plant by the xylem vessels. Sugar made in the leaves is carried to every part of a plant by the phloem.

As water is lost through the stomata in the leaves more is taken in by the roots. This continuous flow of water from the roots to the leaves is called the transpiration stream. If water is lost from the leaves faster than it is taken in by the roots the plant wilts.

Key words

phloem	Carries sugar from the leaves to other parts of the plant
transpiration	Loss of water through the stomata in the leaves
vascular bundles	Made up of xylem and phloem. They carry substances from one part of a plant to another
xylem	Carries water and minerals from the roots to other parts of the plant

Questions

1 Describe how root hairs take in water.

2 This is a cross section of a stem. Name the parts labelled A, B and C

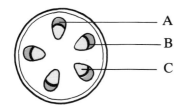

3 This experiment shows how the weather affects the amount of water lost from a plant by transpiration.

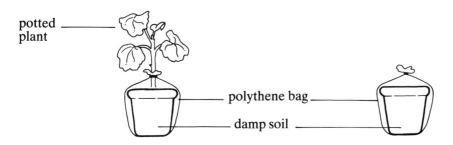

Both pots were weighed each day for nine days. The results are shown below:

days from the start of the experiment	1	2	3	4	5	6	7	8	9
weather	mild	cloudy	windy warm	mild	sunny	sunny	sunny	mild	windy mild
loss of weight of pot with plant	14 g	10 g	27 g	13 g	20 g	22 g	18 g	14 g	16 g
loss of weight of pot without plant	0 g	0 g	0 g	0 g	0 g	0 g	0 g	0 g	0 g

(a) Why were the pots covered with polythene bags?
(b) Which day did the pot with a plant lose most weight?
(c) What was the weather like that day?
(d) Which day did the pot with a plant lose least weight?
(e) What was the weather like that day?
(f) What was the total weight lost by the pot with a plant in this experiment?
(g) Does the weather affect the amount of water lost by a plant? Explain your answer.
(h) Explain why the pot without a plant stayed the same weight for the nine days.

4 In an experiment four leaves were cut from a plant. Different parts of each leaf were covered with vaseline (a waterproof jelly). The leaves were hung on a line and left for 2 weeks.

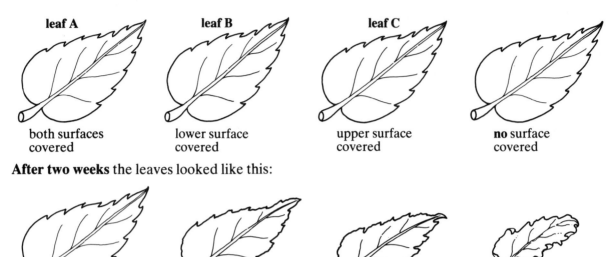

leaf A	**leaf B**	**leaf C**	
both surfaces covered	lower surface covered	upper surface covered	**no** surface covered

After two weeks the leaves looked like this:

(a) Which leaf lost the most water? Give reasons for your answer.
(b) Which leaf lost the least water? Explain your answer.
(c) Is more water lost through (1) the upper surface or (2) the lower surface of a leaf? Give reasons for your answer.

5 When a root tip is cut in half you can see the following parts:

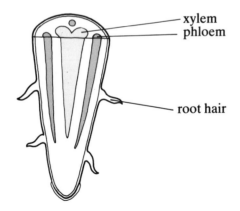

xylem
phloem

root hair

Can you find the same parts labelled X Y and Z on this section across a root tip?

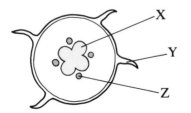

X
Y
Z

12: Transport in animals

Digested foods, oxygen and waste substances must be moved quickly from one part of an animal to another. In a unicellular animal like Amoeba these substances can move from one part of the cell to another by diffusion (see page 9). In a large multicellular animal it would take too long for these substances to diffuse to every cell in the body. These substances are carried from one part of the body to another by the **blood**.

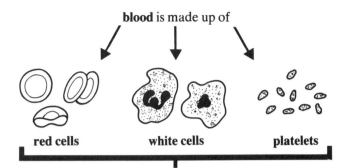

blood is made up of

red cells   white cells   platelets

floating in a yellow watery liquid called **plasma**

Red cells

Red cells do not have a nucleus. They are red because they have **haemoglobin** in them. The haemoglobin carries oxygen round the body.

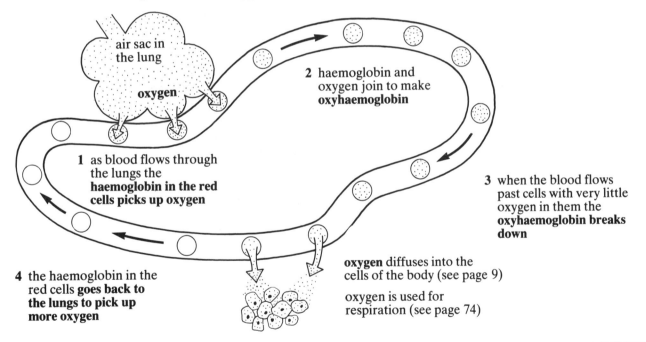

air sac in the lung

oxygen

2 haemoglobin and oxygen join to make **oxyhaemoglobin**

1 as blood flows through the lungs the **haemoglobin in the red cells picks up oxygen**

3 when the blood flows past cells with very little oxygen in them the **oxyhaemoglobin breaks down**

4 the haemoglobin in the red cells **goes back to the lungs to pick up more oxygen**

oxygen diffuses into the cells of the body (see page 9)

oxygen is used for respiration (see page 74)

Questions

1 Explain why a unicellular organism like Amoeba does not need blood.
2 What is the liquid part of blood called?
3 Which blood cells have haemoglobin in them?
4 What is oxyhaemoglobin?

White cells

Some white cells can 'eat' the bacteria which cause disease. These white cells are called **phagocytes**.

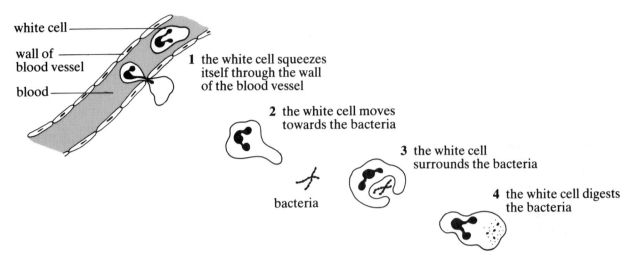

white cell

wall of blood vessel

blood

1 the white cell squeezes itself through the wall of the blood vessel

2 the white cell moves towards the bacteria

bacteria

3 the white cell surrounds the bacteria

4 the white cell digests the bacteria

Platelets

Platelets are made in the bone marrow. They help the blood to clot.

When the skin is cut:

1 Bleeding cleans the cut.

blood

skin

2 Threads of fibrin are made.
When the platelets come in contact with air they make a chemical which changes a protein in the blood into threads of fibrin.

threads of fibrin

3 Red cells get trapped in the threads of fibrin and make a **clot**.

red cells trapped in fibrin

4 The blood clot:
(a) stops the bleeding
(b) seals the cut to stop dirt and bacteria getting into the body

blood clot

5 As the clot dries it goes hard and makes a **scab**. New skin is made underneath the scab.

Questions

1 What are phagocytes?
2 Why are phagocytes important?
3 Why are platelets important?

4 How is fibrin made?
5 What does a blood clot do?

Functions of the blood

1 The blood carries substances from one part of the body to another.

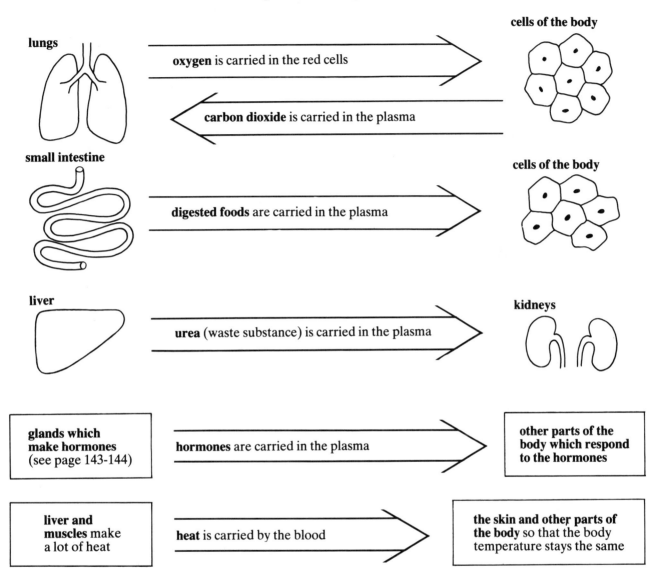

lungs

oxygen is carried in the red cells

carbon dioxide is carried in the plasma

cells of the body

small intestine

digested foods are carried in the plasma

cells of the body

liver

urea (waste substance) is carried in the plasma

kidneys

glands which make hormones (see page 143-144)

hormones are carried in the plasma

other parts of the body which respond to the hormones

liver and muscles make a lot of heat

heat is carried by the blood

the skin and other parts of the body so that the body temperature stays the same

2 The blood helps to prevent disease.

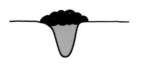

If the skin is cut the **blood clots** to stop dirt and bacteria getting into the body. (see page 94)

Some **white cells surround and digest the bacteria** which get into the body (see page 94).

Some white cells make **antibodies** which help to destroy bacteria and the poisonous substances made by bacteria (toxins). The body makes different antibodies for different diseases.

Immunity

The antibodies made to fight an infection may make the body **immune** to the disease. The next time the body comes into contact with the disease the antibodies can be made quickly. In some cases this stops the body being infected again. In other cases it makes the second attack milder than the first.

Vaccines can be injected into the body to give a mild form of a disease. The body makes antibodies to fight the disease. This makes the body immune to the disease.

Questions

1 Name four substances carried by the blood plasma.
2 Which part of the blood carries oxygen to the cells of the body?
3 What happens to heat made in the liver and muscles?
4 Explain how antibodies help to prevent disease.
5 Why are vaccines injected into the body?

Blood vessels

Blood is carried round the body in **blood vessels**. There are three types of blood vessels. They are called **arteries, veins** and **capillaries**.

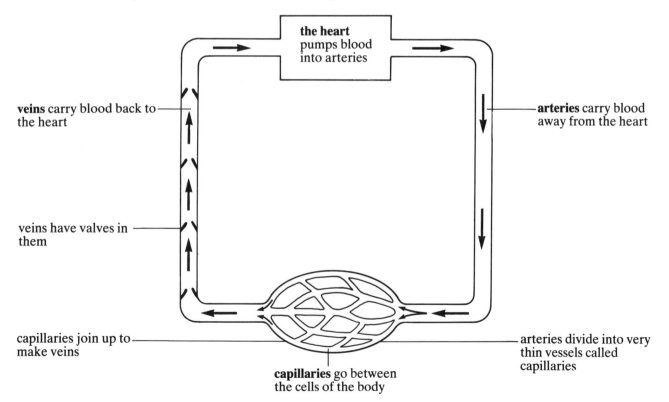

the heart pumps blood into arteries

veins carry blood back to the heart

veins have valves in them

capillaries join up to make veins

capillaries go between the cells of the body

arteries carry blood away from the heart

arteries divide into very thin vessels called capillaries

Arteries

As the blood is pumped along an artery the muscle fibres are stretched. After being stretched the muscle fibres contract and press inwards. This helps to push the blood towards the capillaries. The stretching and contracting in arteries causes the **pulse** which can be felt at the wrist and neck.

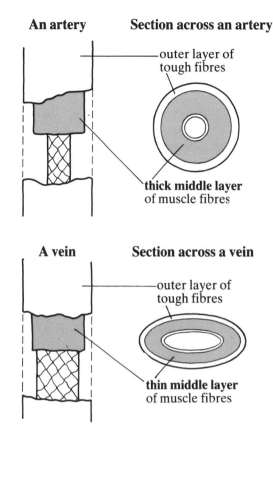

An artery   **Section across an artery**

outer layer of tough fibres

thick middle layer of muscle fibres

Veins

Veins have only a thin layer of muscle fibres. The valves in a vein stop the blood flowing away from the heart.

Valves open like double doors to let the blood flow towards the heart.

Valves close to stop blood flowing away from the heart.

A vein   **Section across a vein**

outer layer of tough fibres

thin middle layer of muscle fibres

Capillaries

Capillaries are very small blood vessels. They have very **thin walls**. Substances can go through the thin walls of a capillary.

A capillary

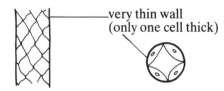

very thin wall (only one cell thick)

Questions

1 Name the three types of blood vessel.
2 Which type of blood vessel has a very thick layer of muscle fibres?
3 Which blood vessels carry blood back to the heart?
4 Which blood vessels go between the cells of the body?
5 Which blood vessels carry blood away from the heart?
6 Where would you feel a pulse?
7 Which blood vessels have valves in them?

The heart

The blood is pumped to all parts of the body by the **heart**. The blood is pumped through two systems:

1 blood is pumped from the heart **to the lungs** and back again

2 blood is pumped from the heart **to all other parts of the body** and back again

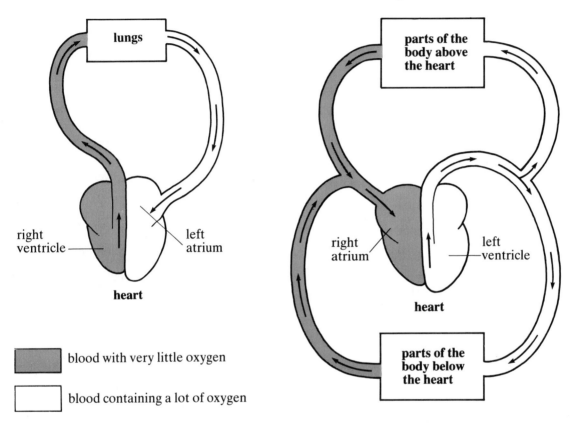

blood with very little oxygen

blood containing a lot of oxygen

The main parts of the heart

The wall of the heart is made of **cardiac muscle cells** (see page 113). The muscle cells must get enough food and oxygen for them to contract and keep the heart beating. The **coronary artery** carries food and oxygen to the heart muscle cells.

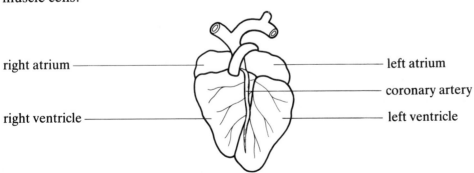

right atrium

right ventricle

left atrium

coronary artery

left ventricle

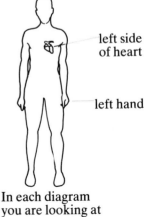

left side of heart

left hand

In each diagram you are looking at the heart in this position

The heart has four main parts or **chambers**. The top chambers are called **atria**. (Each one is called an atrium.) The bottom chambers are called **ventricles**.

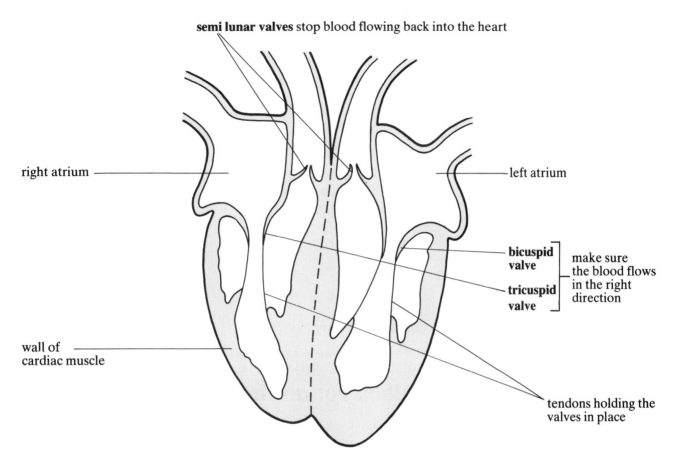

semi lunar valves stop blood flowing back into the heart

right atrium

left atrium

bicuspid valve

tricuspid valve

make sure the blood flows in the right direction

wall of cardiac muscle

tendons holding the valves in place

Questions

1 Name the blood vessel which supplies the muscle of the heart with blood.
2 How many chambers does the heart have?
3 What are the top chambers of the heart called?
4 What are the lower chambers of the heart called?
5 What is another name for heart muscle?
6 Where is the tricuspid valve found?
7 Where is the bicuspid valve found?

How blood flows through the heart

The next diagram is a simple plan of the heart. It shows the main blood vessels and explains how the blood flows through the heart.

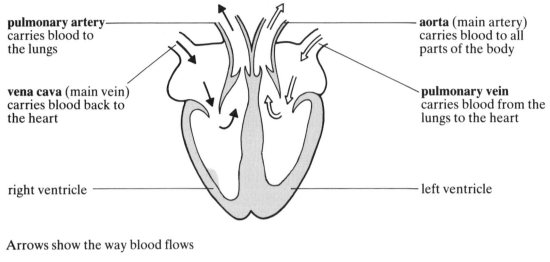

pulmonary artery carries blood to the lungs

vena cava (main vein) carries blood back to the heart

aorta (main artery) carries blood to all parts of the body

pulmonary vein carries blood from the lungs to the heart

right ventricle

left ventricle

Arrows show the way blood flows

⟹ blood carrying a lot of oxygen (**oxygenated** blood)

⟶ blood carrying a lot of carbon dioxide (**deoxygenated** blood)

How the heart pumps blood round the body

1 When the muscles of the **ventricles relax** the blood flows into the ventricles.

2 When the muscle of the **ventricles contract** the blood is pumped out of the heart.

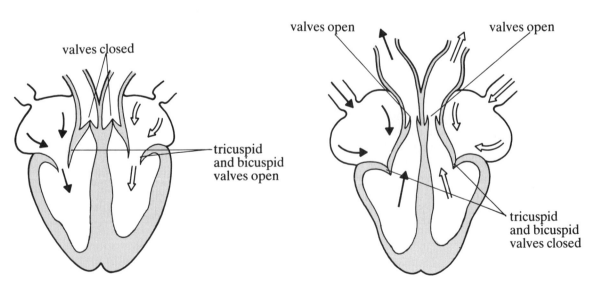

valves closed

tricuspid and bicuspid valves open

valves open

valves open

tricuspid and bicuspid valves closed

A **heart beat** is one contraction and relaxation of the heart. In man, the heart normally beats 60-80 times a minute. During exercise the heart may beat as fast as 150 times a minute.

The circulatory system

Blood **circulates** to every cell in the body. This diagram shows the main blood vessels and explains how the blood circulates to all parts of the body.

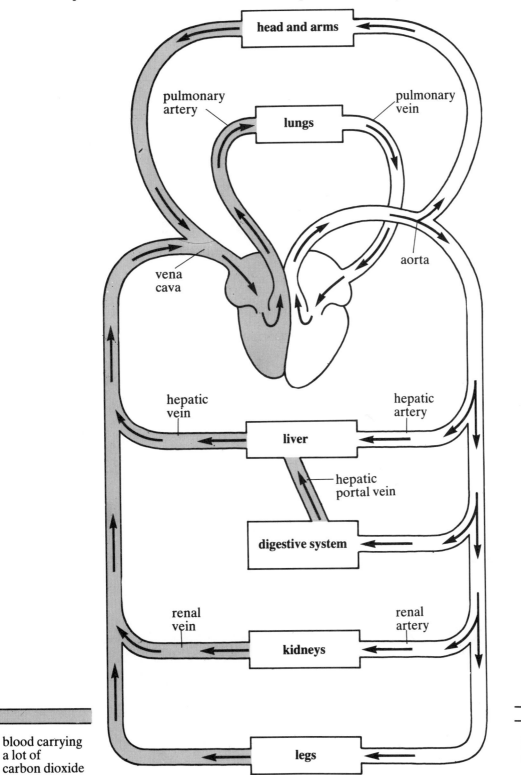

Questions

1 Which blood vessel carries blood to the right atrium of the heart?
2 **(a)** Name the blood vessel which carries blood from the right ventricle of the heart to the lungs.
   (b) Is the blood oxygenated or deoxygenated?
3 Which vein carries blood from the lungs to the heart?
4 What is oxygenated blood?
5 Name the blood vessel which carries blood to the liver.
6 Which blood vessel carries blood from the digestive system to the liver?
7 Which blood vessels carry oxygenated blood from the heart to the kidneys?

How substances pass between a capillary and the cells

Capillary walls are so thin that liquid from the blood leaks out into the spaces between the cells. This liquid is called **tissue fluid**. The tissue fluid carries oxygen, digested foods and other useful substances from the blood to the cells. The tissue fluid also carries carbon dioxide, urea and other waste substances away from the cells. After picking up waste substances, some of the tissue fluid goes back into the capillary and becomes part of the blood. Most of the tissue fluid goes into a **lymph vessel** and becomes the **lymph**.

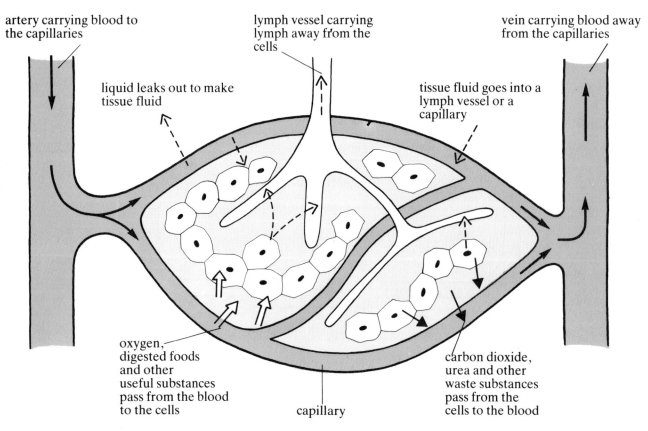

artery carrying blood to the capillaries

lymph vessel carrying lymph away from the cells

vein carrying blood away from the capillaries

liquid leaks out to make tissue fluid

tissue fluid goes into a lymph vessel or a capillary

oxygen, digested foods and other useful substances pass from the blood to the cells

capillary

carbon dioxide, urea and other waste substances pass from the cells to the blood

The lymph system

Lymph vessels carry the lymph to a large vein just above the heart. The lymph then goes into the blood system. In this way the liquid lost from the capillaries is returned to the blood. This diagram shows how the lymph from the left arm gets back into the blood.

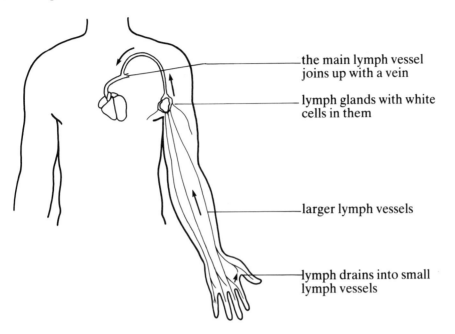

the main lymph vessel joins up with a vein

lymph glands with white cells in them

larger lymph vessels

lymph drains into small lymph vessels

The lymph system is important because it:

1 returns liquid to the blood.
2 helps to prevent disease. The white cells in the lymph glands destroy bacteria and other things which could cause disease. Lymph glands are found in the neck, armpit, groin and small intestine.
3 carries fats away from the villi in the small intestine (see page 60).

Questions

1 Where is tissue fluid found?
2 Why is tissue fluid important?
3 Name two substances carried to the cells.
4 Name two substances carried away from the cells.
5 Explain why the lymph system is important.
6 Where, in the body, would you find lymph glands?
7 Why are lymph glands important?

Summary

Blood carries substances from one part of the body to another. The blood also prevents disease by clotting, by destroying bacteria and by making antibodies. The blood circulates to every cell in the body in a system of blood vessels. Arteries carry blood away from the heart. Arteries divide into capillaries which go in between the cells of the body. The capillaries join up to make veins which carry the blood back to the heart.

Tissue fluid carries substances between a capillary and the cells. The tissue fluid may become part of the blood or drain into a lymph vessel and become lymph. The lymph eventually goes into the blood system.

Key words

antibodies	Chemicals made by the blood to destroy harmful bacteria
artery	A blood vessel which carries blood away from the heart
capillary	A small blood vessel which carries blood in between the cells
haemoglobin	The red substance in red blood cells. Carries oxygen to the cells of the body
immune	When the body is not affected by a disease
lymph gland	Lymph glands have white cells in them which remove bacteria and dead cells from the lymph.
lymph system	A network of tubes which carries lymph from the cells of the body to the blood system
plasma	The liquid part of blood
vein	A blood vessel which carries blood towards the heart

Questions

1 Give four functions of the blood.

2 Copy and complete this paragraph.

Blood leaves the right............ of the heart and goes along the pulmonary............ to the lungs. In the lungs the red pigment............ picks up............ to make oxyhaemoglobin.

3 Name the waste substance carried by the blood, from the liver to the kidneys.

4 Copy and complete this table to show two more differences between arteries and veins.

	arteries	veins
1	have a thick middle layer	have a thin middle layer
2		
3		

5 Explain how food absorbed in the small intestine reaches the blood stream.

6 Draw labelled diagrams of **(a)** a red blood cell and **(b)** a white blood cell. Give one function of each type of cell.

7 Explain how substances pass between a capillary and the cells of the body.

8 How does an injection of vaccine protect a person against disease?

9 Explain how the blood helps to prevent disease.

10 The diagram shows a mammalian heart

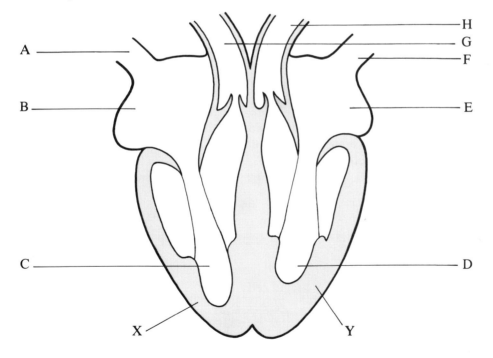

(a) Name the chambers B C D and E.
(b) Name the blood vessels A F G and H.
(c) Why is the muscle in the heart wall thinner at X than at Y?
(d) Which side of the heart is filled with oxygenated blood?
(e) What is the normal 'heart beat' in man?
(f) Why do you think the heart beats faster during exercise?

13: Support and movement

All plants and animals move in some way. Some parts of a plant move very slowly. The petals of a flower open when the sun shines. The leaves of a plant spread out to trap the light needed to make food. Most animals can move from one place to another. They move quickly to catch their food or to get away from their enemies.

Plants and animals must be supported so that they can move and keep their shape. Plants and animals living in the sea, lakes or ponds are supported by the water around them. Plants and animals living on land have special parts of the body which give support, but do not stop movement.

Support in plants

Cells which are full of water help to support a plant. Imagine this plastic bottle is a plant cell.

When the bottle is full of water it feels firm.

When there is only a little water in the bottle it feels soft and squashes easily.

When the bottle is full the water pushes outwards against the sides of the bottle. This makes the whole bottle feel firm. Water pushing outwards against the cellulose cell wall makes a plant cell firm or **turgid**.

Turgor

A plant with turgid cells is firm and strong.

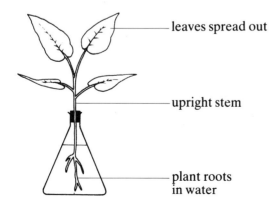

leaves spread out

upright stem

plant roots in water

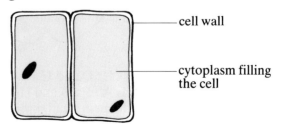

turgid cells

cell wall

cytoplasm filling the cell

Wilting

A plant loses water through its leaves (see page 89). If water is lost faster than it is taken in by the roots, the cells lose some of their water. The leaves and stem become limp and the plant **wilts**.

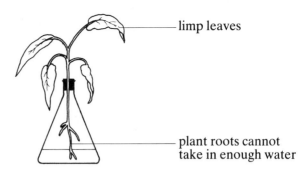

limp leaves

plant roots cannot take in enough water

Plasmolysis

If plant roots are put into a strong sugar solution water is drawn out of the cells by osmosis (see page 11). The plant cells lose so much water that they become **plasmolysed**.

Plasmolysed cells

limp stem

limp leaves

plant roots in a strong sugar solution

cell wall

cytoplasm shrunk away from cell wall

Xylem vessels

The xylem vessels make a woody 'skeleton'. The xylem vessels help to support the softer parts of the roots, stem and leaves.

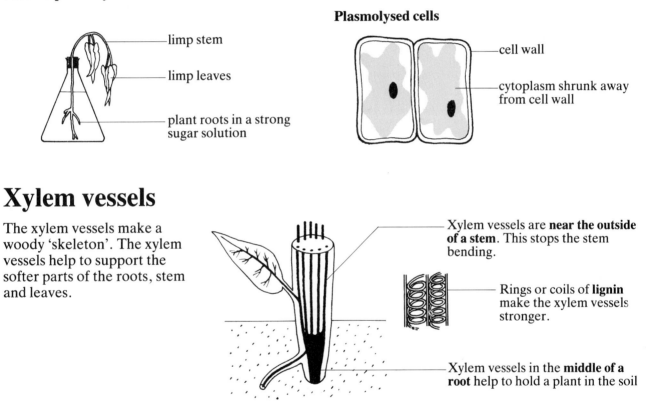

Xylem vessels are **near the outside of a stem**. This stops the stem bending.

Rings or coils of **lignin** make the xylem vessels stronger.

Xylem vessels in the **middle of a root** help to hold a plant in the soil

---

Questions

1 Explain why plants and animals must be supported in some way.
2 What are turgid cells like?
3 How would a plant look if its cells were turgid?

4 What happens to the cytoplasm when a cell becomes plasmolysed?
5 Which part of a plant makes a woody 'skeleton'?
6 Give two examples of plant movements.

Support in animals

Some animals are supported by a **skeleton**. There are two main types of skeleton:

1 A skeleton outside the body is called an **exoskeleton**. Crabs, lobsters and insects have an exoskeleton.
2 A skeleton inside the body is called an **endoskeleton**. Fish, amphibians, reptiles, birds and mammals have an endoskeleton.

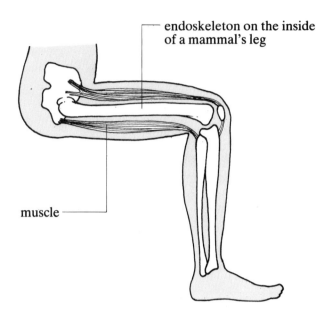

endoskeleton on the inside of a mammal's leg

muscle

exoskeleton on the outside of an insect's leg

muscle

The locust – an example of an exoskeleton

The locust has a hard, strong **cuticle** on the outside of its body.

The cuticle:

1 **supports** the body
2 **protects** the soft parts inside the body
3 stops the body drying out
4 stops dirt and bacteria getting into the body

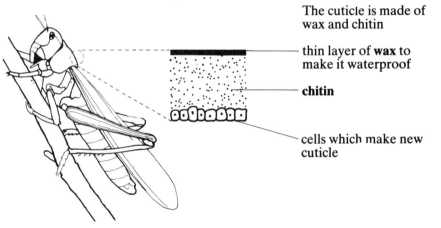

The cuticle is made of wax and chitin

thin layer of **wax** to make it waterproof

chitin

cells which make new cuticle

Questions

1 What is an exoskeleton?
2 Name three animals which have an exoskeleton.
3 What is an endoskeleton?
4 Which groups of animals have endoskeletons?
5 What makes the locust's cuticle waterproot?
6 Explain why a locust's cuticle is important.

Man – an example of an endoskeleton

The skeleton of man is important because it:

1 **supports** the body
2 **protects** some organs of the body
3 helps a person **move**

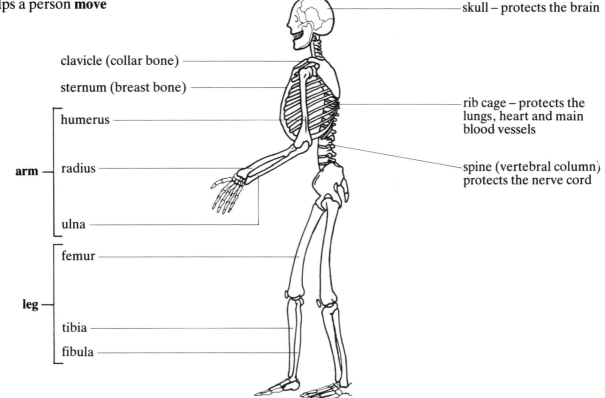

skull – protects the brain

clavicle (collar bone)

sternum (breast bone)

rib cage – protects the lungs, heart and main blood vessels

spine (vertebral column) protects the nerve cord

arm
- humerus
- radius
- ulna

leg
- femur
- tibia
- fibula

Bone

The skeleton is made of **bone**.

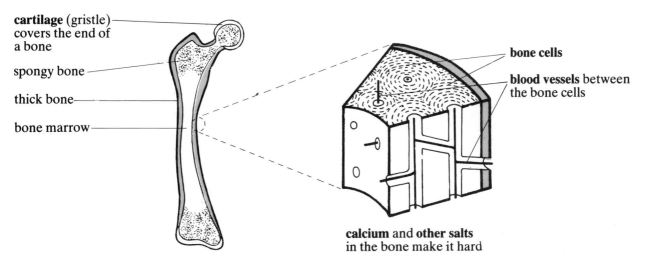

cartilage (gristle) covers the end of a bone

spongy bone

thick bone

bone marrow

bone cells

blood vessels between the bone cells

calcium and **other salts** in the bone make it hard

The spine

The spine:

1 is **curved**

2 is made up of many small bones called **vertebrae**

3 **can bend** backwards and forwards as well as from side to side

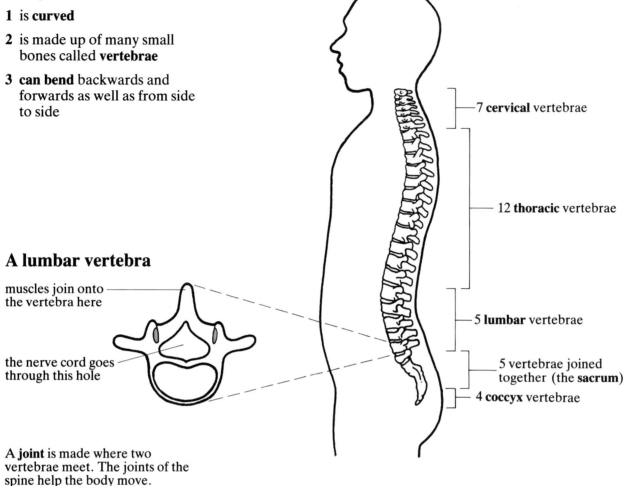

7 **cervical** vertebrae

12 **thoracic** vertebrae

5 **lumbar** vertebrae

5 vertebrae joined together (the **sacrum**)

4 **coccyx** vertebrae

A lumbar vertebra

muscles join onto the vertebra here

the nerve cord goes through this hole

A **joint** is made where two vertebrae meet. The joints of the spine help the body move.

There is a pad of cartilage (intervertebral disc) between the vertebrae.

Questions

1 Why is the skeleton of man important?
2 Why are calcium salts an important part of a bone?
3 What is cartilage?
4 Explain how the skeleton protects some organs of the body.
5 Name the three bones of the leg.
6 What are vertebrae?
7 What are intervertebral discs made of?

Joints in man

There are three types of joint in the body:

1 Joints which do not move. These are called **immovable joints**. There are immovable joints in the skull.
2 Joints which move a little. These are called **slightly movable joints**. There are slightly movable joints in the spine.
3 Joints which move a lot. These are called **freely movable joints**. The hip, elbow, shoulder and knee are all freely movable joints.

The hip joint

The hip joint is called a **ball and socket joint**. The rounded head (ball) of one bone fits into a socket in another bone.

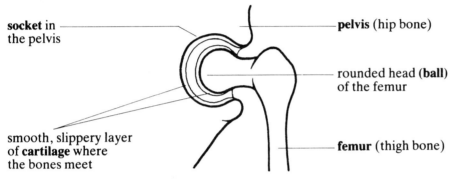

socket in the pelvis

pelvis (hip bone)

rounded head **(ball)** of the femur

smooth, slippery layer of **cartilage** where the bones meet

femur (thigh bone)

A ball and socket joint can move in **all directions**.

The elbow joint

The elbow joint is called a **hinge joint**.

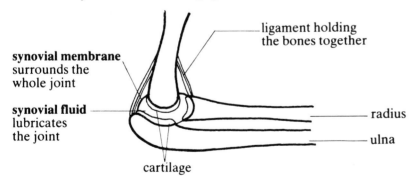

ligament holding the bones together

synovial membrane surrounds the whole joint

synovial fluid lubricates the joint

radius

ulna

cartilage

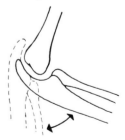

A hinge joint can move in only **one direction**.

Questions

1 Where would you find **(a)** immovable joints and **(b)** slightly movable joints in the body?

2 What are ligaments?
3 Which type of joint can move **(a)** in all directions **(b)** in only one direction?

Muscles and muscle action

Muscles can **contract** or get shorter. Muscles usually work in pairs. A pair of muscles pulls a joint in opposite directions. These diagrams show how a pair of muscles in the arm pull the elbow joint in different directions.

When the **biceps muscle** contracts the arm bends.

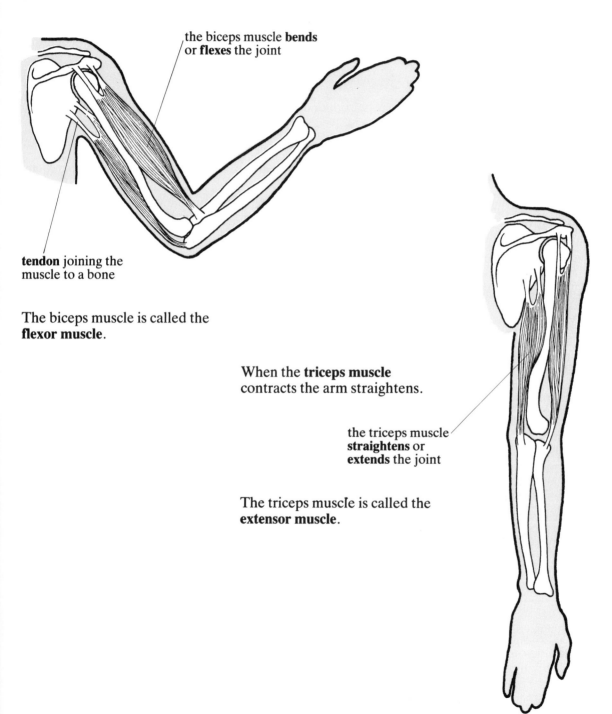

the biceps muscle **bends** or **flexes** the joint

tendon joining the muscle to a bone

The biceps muscle is called the **flexor muscle**.

When the **triceps muscle** contracts the arm straightens.

the triceps muscle **straightens** or **extends** the joint

The triceps muscle is called the **extensor muscle**.

Types of muscle in man

There are three types of muscle in the body:

1 **striped muscle**
2 **smooth muscle**
3 **cardiac** (heart) **muscle**

Striped muscle

Muscles which move joints are made up of
striped muscle fibres. These fibres can contract
quickly but they soon get tired. A person can
consciously control the contraction so it is also
called **voluntary muscle**.

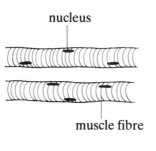

nucleus

muscle fibre

Smooth muscle

Smooth muscle is found in the wall of the gut and
the bladder. Smooth muscle contracts slowly. A
person cannot consciously control the
contraction so it is called **involuntary muscle**.

nucleus

Cardiac muscle

Cardiac muscle is found in the wall of the heart.
Cardiac muscle contracts quickly and
rhythmically throughout life. It is an **involuntary
muscle**.

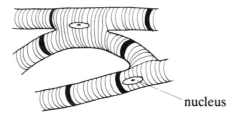

nucleus

Questions

1 Which muscle bends or flexes the arm?
2 Which muscle straightens or extends the arm?
3 Does a muscle get longer or shorter when it contracts?
4 Which type of muscle moves the joints in the body?
5 Name two types of involuntary muscle.

How a fish moves

A fish has a **streamlined shape**. All the fins can lie flat against the body to make the fish more streamlined when it swims.

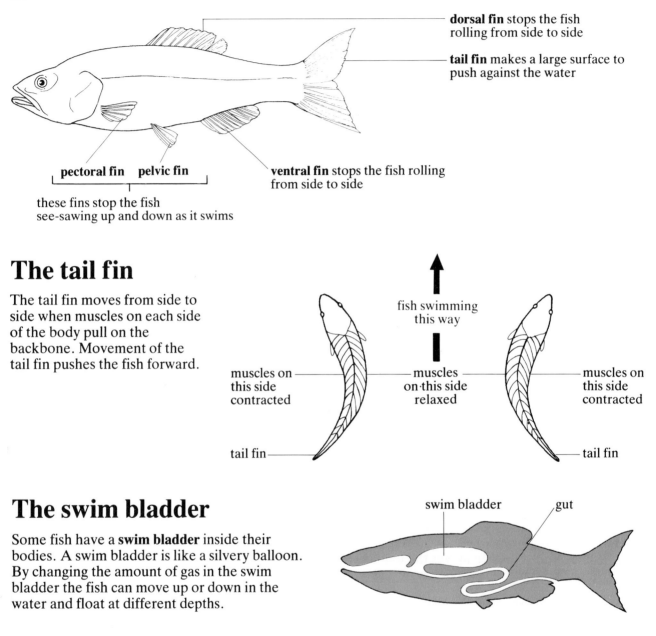

dorsal fin stops the fish rolling from side to side

tail fin makes a large surface to push against the water

pectoral fin   **pelvic fin**

these fins stop the fish see-sawing up and down as it swims

ventral fin stops the fish rolling from side to side

The tail fin

The tail fin moves from side to side when muscles on each side of the body pull on the backbone. Movement of the tail fin pushes the fish forward.

fish swimming this way

muscles on this side contracted

muscles on·this side relaxed

muscles on this side contracted

tail fin

tail fin

The swim bladder

Some fish have a **swim bladder** inside their bodies. A swim bladder is like a silvery balloon. By changing the amount of gas in the swim bladder the fish can move up or down in the water and float at different depths.

swim bladder

gut

Questions

1 Which fins **(a)** stop the fish rolling from side to side **(b)** stop the fish going up and down **(c)** push the fish forward?
2 Explain how the tail fin moves.
3 Why is the swim bladder important?

How a bird moves

A bird uses its **wings** to fly. The wings have a **large surface** to push against the air.

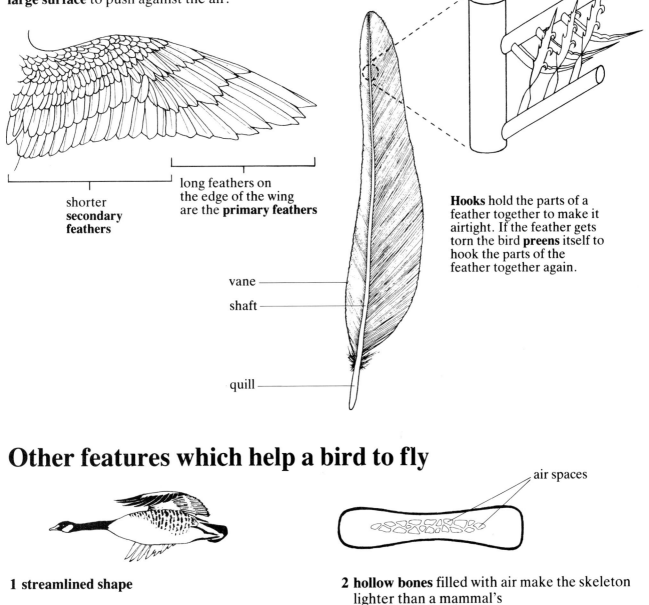

shorter **secondary feathers**

long feathers on the edge of the wing are the **primary feathers**

vane

shaft

quill

Hooks hold the parts of a feather together to make it airtight. If the feather gets torn the bird **preens** itself to hook the parts of the feather together again.

Other features which help a bird to fly

air spaces

1 streamlined shape

2 hollow bones filled with air make the skeleton lighter than a mammal's

this set of muscles pulls the wings up

this set of muscles pulls the wings down

3 strong breast muscles

115

Birds' feet

Some birds use their **feet** to move from place to place.

duck

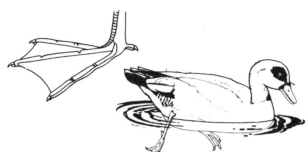

woodpecker

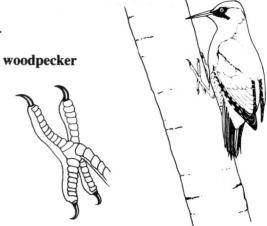

A **webbed foot** has a **large surface** to push against the water.

The woodpecker's **toes** can **grip** the bark of a tree trunk.

Questions

1 Why does a bird's wing have a large surface?
2 What happens if a feather gets torn?
3 Why does a bird need strong breast muscles?
4 Why does a bird have hollow bones?
5 Explain why **(a)** a duck has a webbed foot and **(b)** a woodpecker has two toes pointing forward and two toes pointing backwards.

Summary

Plants and animals must be supported so that they can keep their shape and move. A plant is supported by turgid cells (cells which are full of water) and xylem vessels. The whole plant wilts if there is not enough water in the cells.

Some animals have a skeleton which supports and protects the body. An animal may have either an exoskeleton outside the body or an endoskeleton inside the body.

Man has an endoskeleton made of bone. Where two bones meet a joint is made. Joints are either immovable, slightly movable or freely movable. At a joint the bones are covered with a smooth layer of cartilage and surrounded by a membrane. A fluid inside the membrane lubricates the joint and acts like a cushion or shock absorber.

Joints are moved by muscles which work in pairs. When one muscle contracts the other is stretched.

Key words

cartilage	Smooth, slippery layer which stops bones rubbing together at a joint
cuticle	Strong, hard layer on the outside of an insect. The cuticle is the exoskeleton of an insect
endoskeleton	Skeleton on the inside of an animal
exoskeleton	Skeleton on the outside of an animal
ligament	Tough fibres which hold bones together at joints
plasmolysed cell	A plant cell which has lost most of its water. The cytoplasm shrinks away from the cell wall in a plasmolysed cell
swim bladder	Found in some fish. It is a sac which can be filled with gas to let the fish float at different depths
tendon	Strong fibres which join a muscle to a bone
turgid cell	A plant cell which is swollen with water. The cytoplasm fills the whole cell and pushes outwards against the cell wall to make the cell firm or turgid
vertebra	One of the bones of the spine. Many are called vertebrae

Questions

1 Give three functions of the skeleton in man.

2 Give one example of each of the following types of joint:
(a) immovable, (b) slightly movable, (c) freely movable.

3 Which parts of the body are protected by (a) the skull (b) the rib cage (c) the vertebral column (spine)?

4 This diagram shows the main parts of a freely movable joint.

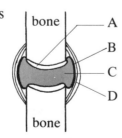

(a) Name the parts labelled A B C and D.
(b) Explain what the parts labelled C and D do.

5 Explain the advantages of having a backbone made up of many separate vertebra.

6 Give two differences between an endoskeleton and an exoskeleton.

7 (a) Name three types of muscle found in the human body.
(b) Say where each type is found.
(c) Explain the difference between voluntary and involuntary muscles.

8 Give two differences between plant and animal movement.

9 Why do birds have **(a)** hollow bones **(b)** very large muscles attached to the breast bone **(c)** large wings?

10 Look at the diagrams below. Which foot belongs to a **(a)** duck **(b)** bird which perches on a branch **(c)** bird which uses its claws to catch other animals **(d)** bird which climbs trees?

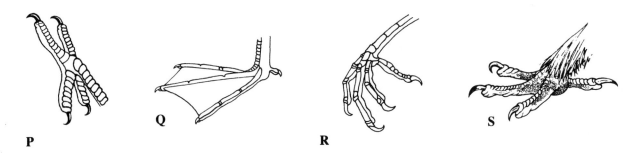

P   Q   R   S

11 Explain how the following parts of a fish help it to move:
(a) fins **(b)** muscles attached to the backbone **(c)** swim bladder.

12 Look at this diagram of a rabbit skeleton. Name the parts labelled A to G.

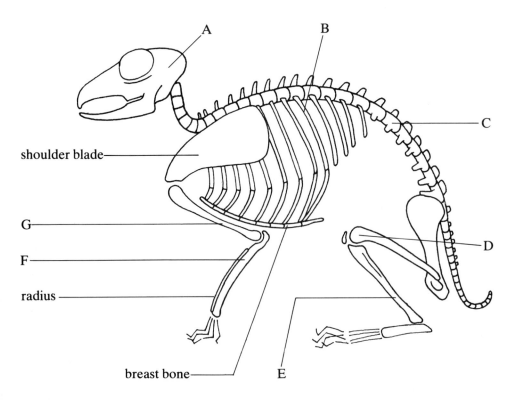

118

14: Excretion

All living cells produce waste substances. The waste substances produced by living cells are called **excretions**. Some excretions are poisonous so they must be removed from the body. Removing waste substances from the body is called excretion.

Excretion in plants

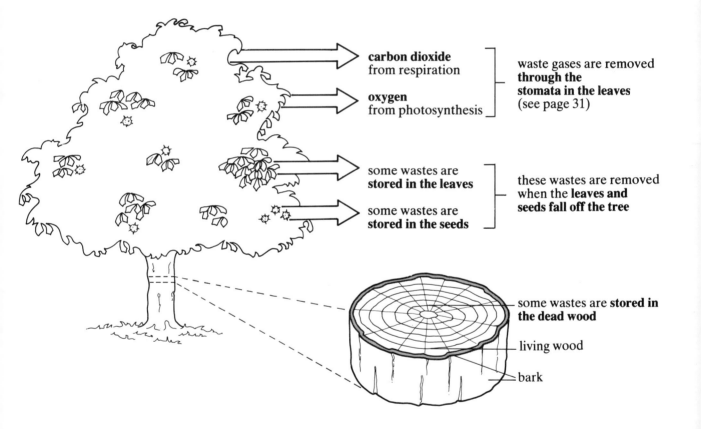

carbon dioxide
from respiration

oxygen
from photosynthesis

waste gases are removed
**through the
stomata in the leaves**
(see page 31)

some wastes are
stored in the leaves

some wastes are
stored in the seeds

these wastes are removed
when the **leaves and
seeds fall off the tree**

some wastes are **stored in
the dead wood**

living wood

bark

Questions

1 Name two waste gases produced by a plant.
2 How are these gases removed from the plant?
3 (a) Name three parts of a tree which can store waste substances.
   (b) Explain how a tree removes these waste substances.

Excretion in man

Waste substances made in the cells are excreted from the body. Undigested food is **egested** from the body. The next diagram explains the difference between excretion and egestion.

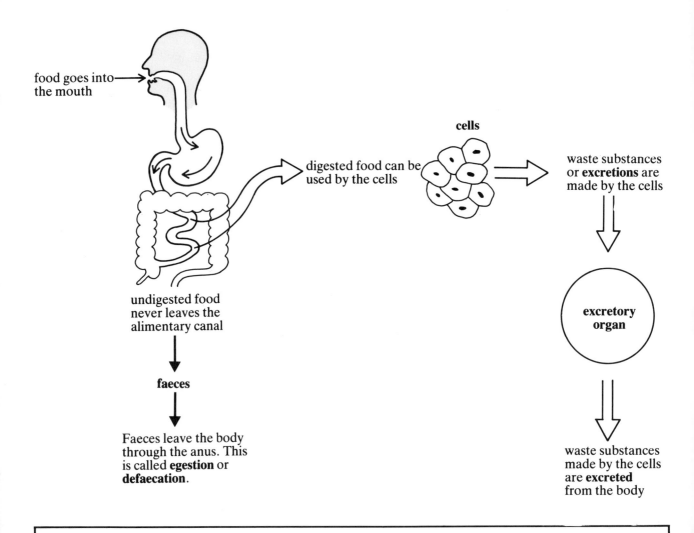

food goes into the mouth

digested food can be used by the cells

cells

waste substances or **excretions** are made by the cells

undigested food never leaves the alimentary canal

faeces

Faeces leave the body through the anus. This is called **egestion** or **defaecation**.

excretory organ

waste substances made by the cells are **excreted** from the body

Questions

1 Explain the difference between egestion and excretion.
2 Explain what happens to undigested food.
3 Where are excretions made?
4 What happens to the excretions?

How carbon dioxide is excreted

When cells respire they produce carbon dioxide.

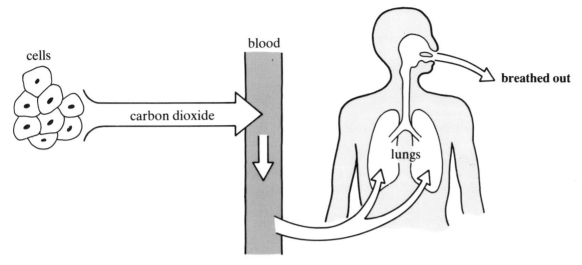

How waste water is excreted

Water is produced by the cells when they respire. Water is also taken into the body with the food we eat. Some of this water must be excreted.

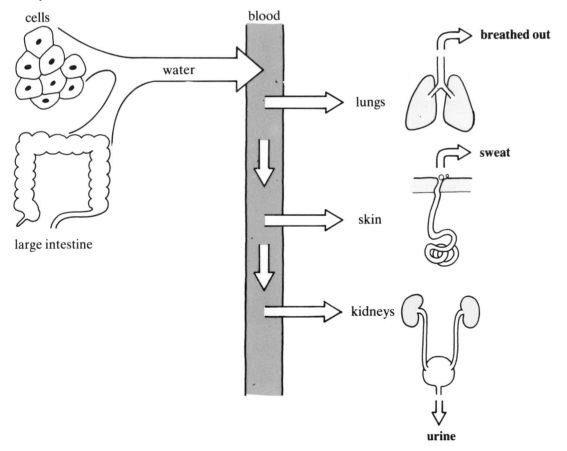

How waste amino acids are excreted

Amino acids which are not needed by the body are changed into **urea**. The urea is excreted from the body in a liquid called **urine**.

How urea is made

1 protein is eaten (protein is made up of amino acids)

2 protein is split into amino acids by digestion

3 amino acids are used by the cells to make new proteins

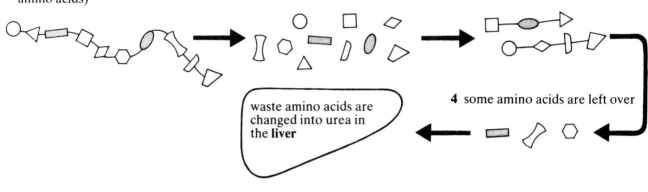

waste amino acids are changed into urea in the **liver**

4 some amino acids are left over

Questions

1 What happens to the carbon dioxide produced by cells of a mammal?

2 Explain how a mammal excretes water.

3 Why do cells need amino acids?

4 What happens to the waste amino acids?

The kidneys

Urine is made in the **kidneys**. The next diagram shows the position of the kidneys in the body.

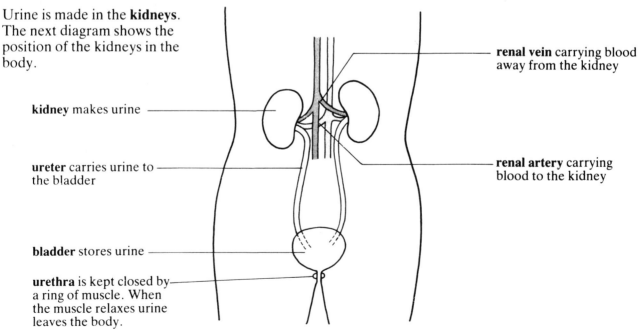

renal vein carrying blood away from the kidney

kidney makes urine

renal artery carrying blood to the kidney

ureter carries urine to the bladder

bladder stores urine

urethra is kept closed by a ring of muscle. When the muscle relaxes urine leaves the body.

Section through a kidney

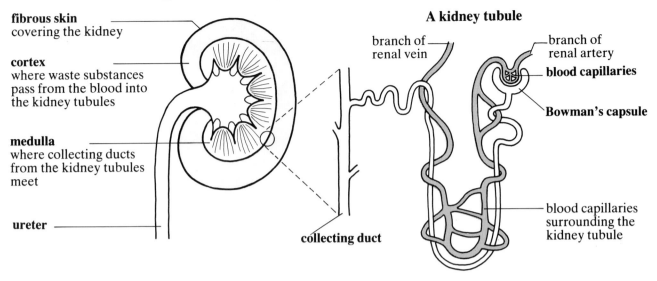

fibrous skin
covering the kidney

cortex
where waste substances
pass from the blood into
the kidney tubules

medulla
where collecting ducts
from the kidney tubules
meet

ureter

collecting duct

A kidney tubule

branch of
renal vein

branch of
renal artery

blood capillaries

Bowman's capsule

blood capillaries
surrounding the
kidney tubule

How a kidney tubule works

The kidney tubules remove urea and other waste substances from the blood.

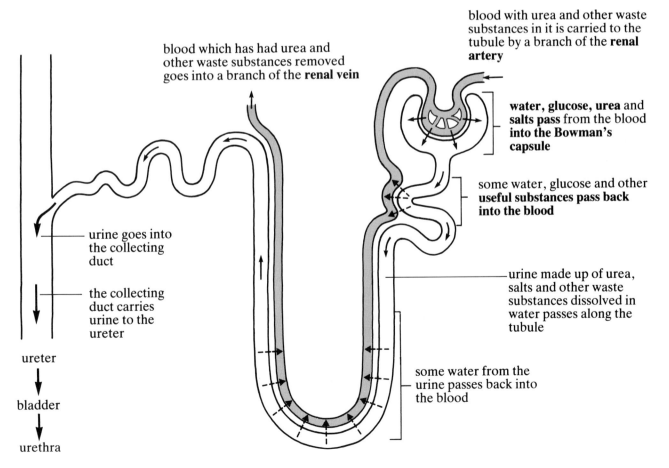

blood which has had urea and
other waste substances removed
goes into a branch of the **renal vein**

blood with urea and other waste
substances in it is carried to the
tubule by a branch of the **renal
artery**

water, glucose, urea and
salts pass from the blood
**into the Bowman's
capsule**

some water, glucose and other
**useful substances pass back
into the blood**

urine goes into
the collecting
duct

the collecting
duct carries
urine to the
ureter

ureter

bladder

urethra

urine made up of urea,
salts and other waste
substances dissolved in
water passes along the
tubule

some water from the
urine passes back into
the blood

Regulation of water by the kidneys

The blood must be at the right **concentration**. It must have enough, but not too much, water in it. If there is **too much water in the blood** a **large volume of dilute urine** is excreted. If there is **not enough water in the blood** a **small volume of concentrated urine** is excreted.

The urine excreted is made up of:

 water

 urea

salts

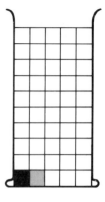

Questions

1 Where is urine made?
2 Which blood vessel carries blood to the kidney?
3 What happens in the cortex of the kidney?
4 Name the substances which pass from the blood into the Bowman's capsule.
5 Which substances pass back into the blood?
6 Which substances dissolve in water to make urine?
7 Explain what happens to the urine made in the kidney tubule.
8 What percentage of urine is water?

How the amount of water in the blood is regulated

Water is lost from the body in the urine. Water is also lost by sweating. There is a balance between the amount of urine leaving the body and the amount of water lost by sweating.

On a cold day

Very little water is lost by sweating.

A lot of urine leaves the body.

On a hot day

A lot of water is lost by sweating.

Very little urine leaves the body.

Sweating

Some **water and salts** are **excreted** through the skin as **sweat**. As sweat dries the body is cooled and waste substances are left on the surface of the skin.

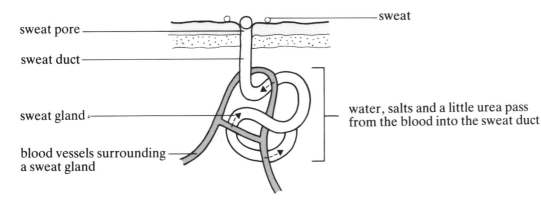

sweat pore

sweat

sweat duct

sweat gland

blood vessels surrounding a sweat gland

water, salts and a little urea pass from the blood into the sweat duct

The skin

This diagram shows the structure of the skin.

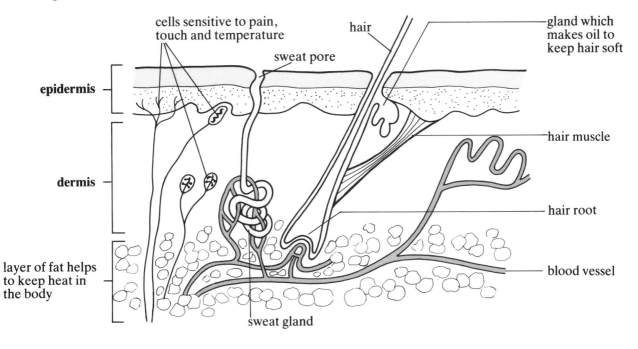

cells sensitive to pain, touch and temperature

hair

gland which makes oil to keep hair soft

sweat pore

epidermis

hair muscle

dermis

hair root

layer of fat helps to keep heat in the body

blood vessel

sweat gland

Questions

1 Explain why more urine leaves the body on a cold day than on a hot day.
2 What happens as sweat dries on the skin?
3 Which waste substances are excreted through the skin?
4 What is the outer layer of skin called?

Functions of the skin

The skin is an important organ which:

1 **excretes water** and **salts**
2 **protects** the body
3 has special cells which are **sensitive to pain, touch** and **temperature**
4 helps to **control the temperature of the body**

Temperature control

When the body is **too hot**:

1 the **blood vessels get larger** so that a lot of
warm blood can be cooled as it flows near the
surface of the skin. This makes the skin look
red.

2 a **lot of sweat** is made

When the body is **too cold**:

1 the **blood vessels contract** so that only a little
warm blood flows near the surface of the skin.
This makes the skin look pale.

2 very **little sweat** is made

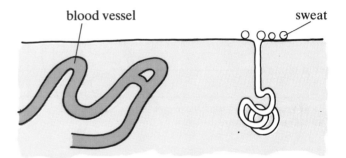

blood vessel   sweat

Animals which live in very cold places have a **thick layer of fat** under the skin.
This helps to keep heat in the body. They also have a **thick layer of hair or
fur**. The air trapped between the hair or fur helps to keep an animal warm.
When an animal is cold the hair muscles contract and the hairs stand up.
This traps more air between the hairs.

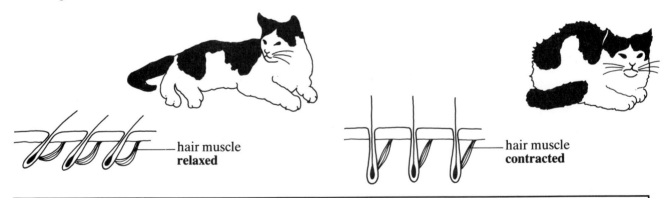

hair muscle **relaxed**   hair muscle **contracted**

Questions

1 Explain why the skin is an important organ.
2 Why do cat's hairs 'stand on end' when it is cold?
3 What happens when a hair muscle relaxes?
4 Why do blood vessels in the skin get larger on a hot day?

Summary

Waste substances must be removed or excreted from the body. A plant removes waste gases through stomata in the leaves. Other waste substances are stored in the leaves, seeds or dead wood so that they do not poison the plant.

The kidneys, the skin and the lungs are the main excretory organs in man. The kidneys remove waste substances such as water, urea and salts from the blood. In this way the kidneys regulate the concentration of the blood.

The skin also helps to control the temperature of the body.

Key words

Bowman's capsule	Cup-shaped part of a kidney tubule
egestion	Undigested food passing out of the anus
excretion	Removing waste substances from the body
urea	A waste substance which is made in the liver and excreted in the urine
ureter	Tube which carries urine from the kidney to the bladder
urethra	Tube which carries urine out of the body
urine	Made from water, urea and salts which have been taken out of the blood by the kidneys

Questions

1 What is excretion?

2 Explain how a plant removes its waste substances.

3 (a) Name three excretory organs in man.
   (b) List the substances excreted by each organ.

4 Draw a large labelled diagram to show the position of the kidneys, ureters and bladder in man.

5 Give four functions of the skin.

6 Explain what happens to amino acids not needed for growth and repair of the body.

7 Describe how urine is made in a kidney tubule.

8 Why does very little urine leave the body if a person has been sweating a lot?

9 Explain how your skin helps to **(a)** lose heat on a hot day **(b)** keep in heat on a cold day.

10 Name the parts of the kidney labelled in this diagram

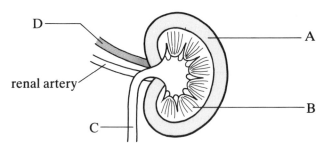

11 Explain what the following parts of the body do:
   (a) kidney   **(b)** ureter   **(c)** bladder   **(d)** urethra

12 Explain why a fox living in the Arctic has a thicker coat than a fox living in Britain.

13 The skin is an excretory organ. Give three other functions of the skin.

14 How do the kidneys regulate the amount of water in the blood?

15 Look at this diagram of the skin
   (a) Name the parts labelled in the diagram.
   (b) Give the functions of the parts labelled A, B, E, G and H.

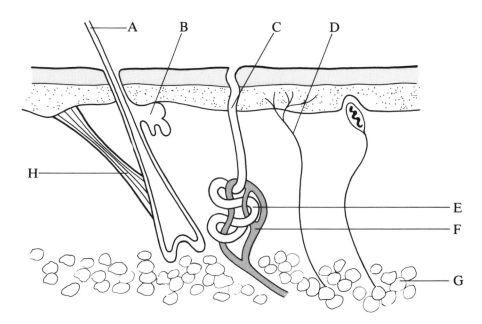

15: Sensitivity

Plants and animals must be aware of what is happening around them. They must be able to **detect any change** in their surroundings. They must be able to **respond to any change**. Any organism which does not detect and respond quickly enough to changes in its surroundings may die.

Sensitivity in plants

Plants **respond slowly** to any change in their surroundings. Plants **respond by growing** in certain directions. Whichever way a seed is planted the shoot grows upwards and the root grows downwards. The shoot grows away from the pull of gravity. The root grows towards the pull of gravity. This response to the pull of gravity is called **geotropism**.

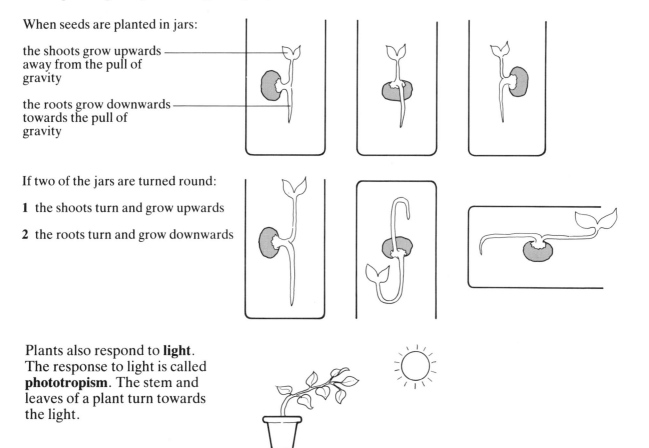

When seeds are planted in jars:

the shoots grow upwards away from the pull of gravity

the roots grow downwards towards the pull of gravity

If two of the jars are turned round:

1 the shoots turn and grow upwards

2 the roots turn and grow downwards

Plants also respond to **light**. The response to light is called **phototropism**. The stem and leaves of a plant turn towards the light.

Geotropism and phototropism help a plant to survive. The roots grow into the soil to get the water and minerals needed by the plant. The leaves turn towards the light which is needed for making food.

How a plant bends towards the light

Phototropism is controlled by a chemical called **auxin**. Auxin is made in the tip of the stem. Auxin makes the plant cells grow faster.

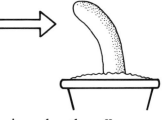
shows where the light is coming from

shows where the auxin is

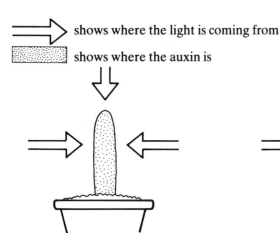

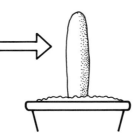

auxin **spread evenly** in the tip of the stem

auxin on the **darker side**

auxin makes the **cells on one side grow faster**. This makes the tip of the stem bend towards the light.

Questions

1 Explain what would happen if you planted a seed 'upside down'.
2 What is geotropism?
3 What is phototropism?

4 How do geotropism and phototropism help a plant to survive?
5 Where is auxin made?
6 What does auxin do?

Sensitivity in animals

Animals respond to **stimuli** such as light, sound, taste, smell, touch and temperature. Cells which are sensitive to one kind of stimulus are grouped together in a **sense organ**.

Sense organs in man

The sense organs in man are the:

Eyes which contain cells sensitive to light.
Ears which contain cells sensitive to sounds and cells which give us a sense of balance.
Tongue which contains cells sensitive to tastes.
Nose which contains cells sensitive to smells.
Skin which contains cells sensitive to touch and temperature.

The eye

The eye is made up of different layers. The layers of the eye are shown in this diagram.

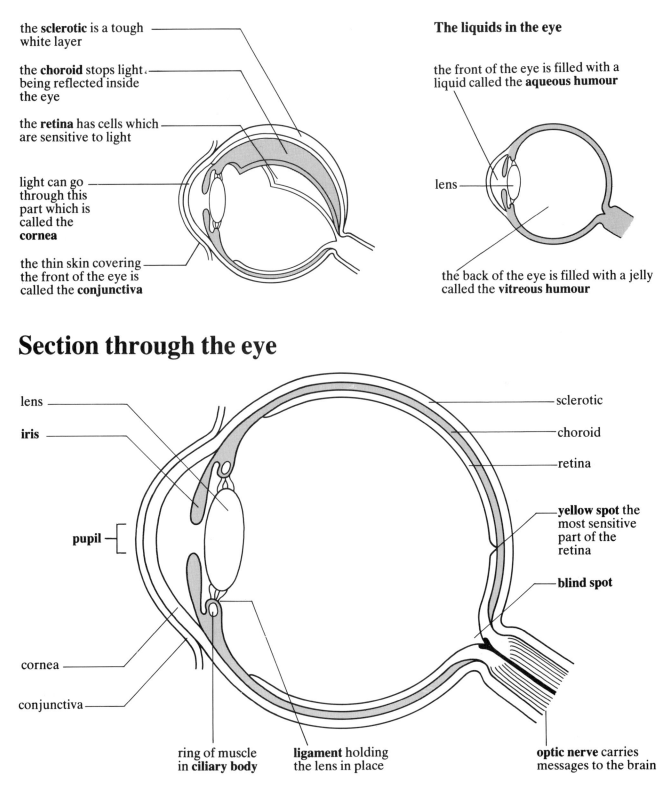

the **sclerotic** is a tough white layer

the **choroid** stops light being reflected inside the eye

the **retina** has cells which are sensitive to light

light can go through this part which is called the **cornea**

the thin skin covering the front of the eye is called the **conjunctiva**

The liquids in the eye

the front of the eye is filled with a liquid called the **aqueous humour**

lens

the back of the eye is filled with a jelly called the **vitreous humour**

Section through the eye

lens

iris

pupil

cornea

conjunctiva

ring of muscle in **ciliary body**

ligament holding the lens in place

sclerotic

choroid

retina

yellow spot the most sensitive part of the retina

blind spot

optic nerve carries messages to the brain

131

The iris and the pupil

The coloured part of the eye is called the **iris**. The iris has two sets of muscles which control the amount of light going into the eye.

In dim light

In bright light

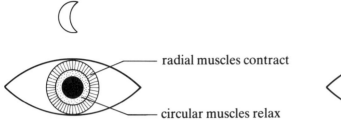

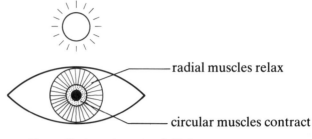

radial muscles contract

circular muscles relax

large pupil lets more light into the eye

radial muscles relax

circular muscles contract

small pupil stops too much light going into the eye

Questions

1 Name the five sense organs in man.
2 What is the outer layer of the eye called?
3 Which layer of the eye has cells which are sensitive to light?
4 What is the conjunctiva?
5 Which part of the eye separates the aqueous humour and the vitreous humour?

6 What does the optic nerve do?
7 What is the coloured part of the eye called?
8 Explain how the iris controls the amount of light going into the eye.

How we see things

Light rays are **bent** as they go through the eye. The light rays are **focused** on the most sensitive part of the retina. The next diagram explains how the eye focuses on an object.

1 light rays bounce off an object

2 the light rays are **bent** as they go through the **cornea** and **lens**

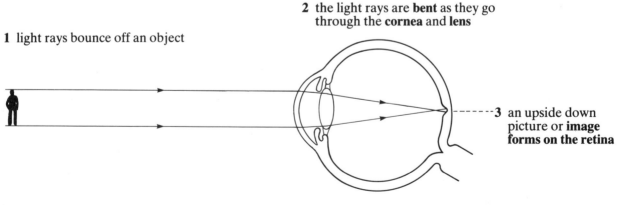

3 an upside down picture or **image forms on the retina**

4 the **optic nerve** carries a message **to the brain**

In the brain the picture is turned up the right way.

How we see things near the eye

The curved surface of the lens helps to focus the light rays on the retina. The lens can change shape to focus the light rays on the most sensitive part of the retina.

If you look at your finger when it is near your eye the things in the distance are out of focus.

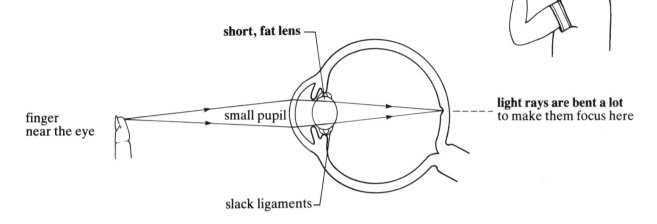

How we see things far away from the eye

If you look at a tree far away from the eye, the things near to it are out of focus.

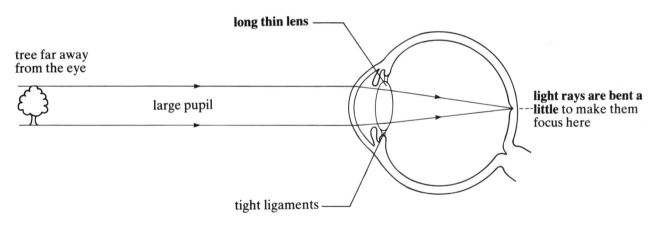

Questions

1 Which parts of the eye bend light rays?
2 (a) What is an image?
  (b) Where is an image formed?
3 Explain what happens when you look at an object near the eye.
4 Why does the lens change shape?

The ear

The ear has three main parts.

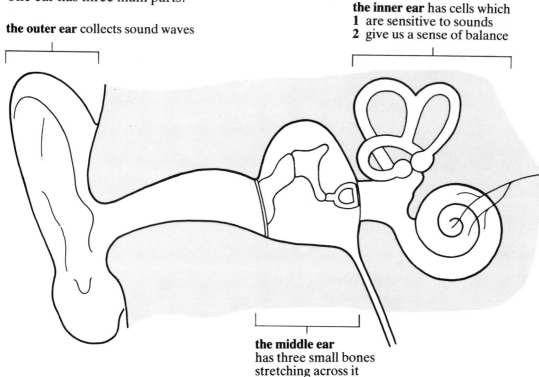

the outer ear collects sound waves

the inner ear has cells which
1 are sensitive to sounds
2 give us a sense of balance

the middle ear
has three small bones
stretching across it

The outer ear

the **pinna** collects
sound waves

auditory canal

ear drum

sound
waves

sound waves make
the ear drum
vibrate in
and out

The middle ear

three small bones called **ossicles**

hammer   anvil   stirrup

oval window

ear drum

vibrations pass from
bone to bone until they
reach the oval window

tube to the throat
(Eustachian tube) lets air
in and out of the middle
ear. This keeps the air
pressure the same on
both sides of the ear
drum.

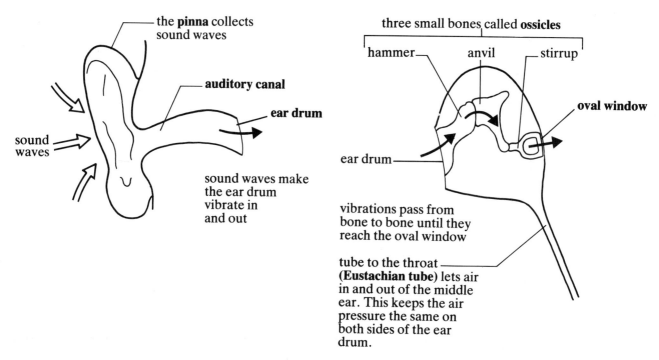

The inner ear

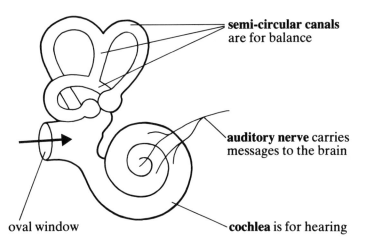

semi-circular canals are for balance

auditory nerve carries messages to the brain

oval window

cochlea is for hearing

Vibrations pass from the oval window to a liquid in the cochlea. Vibrations of the liquid stimulate very small branches of the auditory nerve. The auditory nerve then carries a message to the brain and we hear sounds.

The semi-circular canals

The semi-circular canals also have liquid in them. When the head or body moves, the liquid moves. This stimulates nerve cells and a message is sent to the brain. In this way, a person knows if he is standing on his head or his feet.

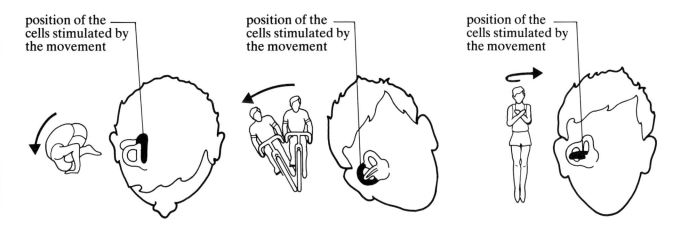

position of the cells stimulated by the movement

position of the cells stimulated by the movement

position of the cells stimulated by the movement

Questions

1 Name the three main parts of the ear.
2 What does the pinna do?
3 What are the ossicles?
4 Which part of the ear helps to keep the air pressure the same on both sides of the ear drum?
5 Explain how vibrations pass from the ear drum to the oval window.
6 Explain what happens in the cochlea.
7 Why are the semi-circular canals important?

The tongue

There are **taste buds** in the tongue. Taste buds have cells which are sensitive to chemicals dissolved in water. There are four types of taste buds in different parts of the tongue. Each type of bud is sensitive to one kind of taste.

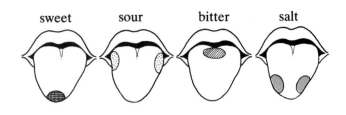

The nose

When a person has a cold, food seems tasteless. This suggests that most of the 'taste' of food is really smell. The nose has **cells which are sensitive to smells** (chemicals in the air). When these cells are stimulated a message is sent to the brain.

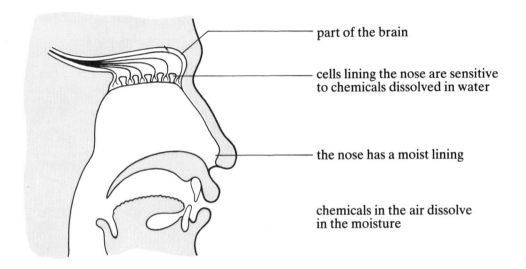

part of the brain

cells lining the nose are sensitive to chemicals dissolved in water

the nose has a moist lining

chemicals in the air dissolve in the moisture

The skin

Some parts of the skin are more sensitive than others. There are a lot of cells sensitive to touch in the fingertips. There are a lot of cells sensitive to pressure on the soles of the feet. The skin also contains cells which are sensitive to pain, heat and cold.

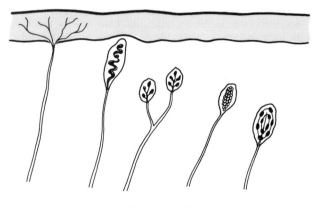

sensory cells in the skin

Summary

A living organism must be able to detect and respond quickly to changes in its surroundings. Plants respond to stimuli such as light and gravity by growing in certain directions. The roots grow downwards in response to the pull of gravity. The shoot responds to light by growing towards it. These responses are known as tropisms or growth movements.

Man has special sense organs to detect changes in his surroundings. The sense organs are the eyes, ears, tongue, nose and skin.

Key words

auxin	A chemical which controls the growth of plant cells
geotropism	A plant's response to the pull of gravity
phototropism	A plant's response to light
response	A change in a plant or animal's behaviour e.g. a plant stem turning towards the light
semi-circular canals	Part of the inner ear concerned with balance
stimulus	Any change in the surroundings which causes a response

Questions

1 What is a plant's response to (a) gravity and (b) light?

2 Give two functions of the ear.

3 Look carefully at the diagram of the eye and
then copy and complete this table.

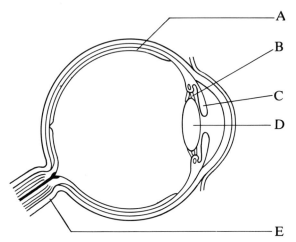

letter on diagram	name of the part	what the part does
A		
B		
C		
D		
E		

4 When they eye is looking at an object near to it the muscles in the ciliary
body contract. What happens to (a) the ligaments holding the lens in place
(b) the shape of the lens?

5 (a) Copy this drawing of the ear. The diagram
is not finished. Draw in and label the:

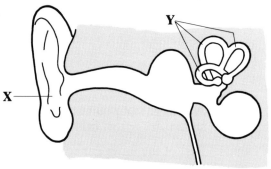

ear drum   ossicles   cochlea   auditory nerve

(b) Name the parts labelled X and Y.
(c) What are the functions of X and Y?

6 Look carefully at this diagram of the tongue.

Which type of taste bud would be stimulated
by (a) sugar solution (b) vinegar (sour taste)
(c) salt solution (d) aspirin (bitter taste)?

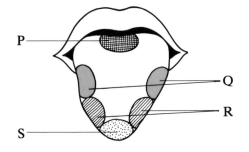

7 If you go into a cinema on a bright sunny day, it takes a few minutes
before you can see the people sitting in the seats. Explain what happens
in the eye during those few minutes.

16: Co-ordination

To stay alive an animal must **respond to stimuli**. The response must be controlled or **co-ordinated** so that different parts of the body work together. The two ways of co-ordinating a response are:

1 by **nerves**

2 by chemicals called **hormones**

The nervous system

The nervous system controls all the organs of the body so that they work together. The main parts of the nervous system are shown in the next diagram.

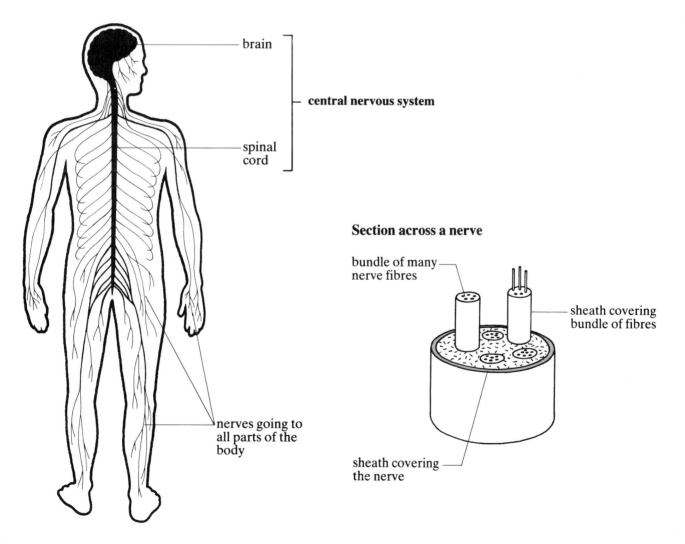

brain

central nervous system

spinal cord

nerves going to all parts of the body

Section across a nerve

bundle of many nerve fibres

sheath covering bundle of fibres

sheath covering the nerve

Nerve cells

The nervous system is made up of nerve cells. There are two types of nerve cells:

1 **sensory** nerve cells

2 **motor** nerve cells

A message called an **impulse** can pass along a nerve fibre. Impulses pass along a nerve fibre in only one direction.

A sensory nerve cell

The arrows show how an impulse passes from a sense organ to the central nervous system.

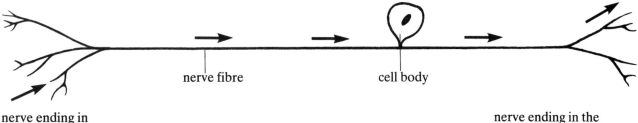

nerve fibre   cell body

nerve ending in
a sense organ

nerve ending in the
central nervous system

A motor nerve cell

The arrows show how an impulse passes from the central nervous system to a muscle.

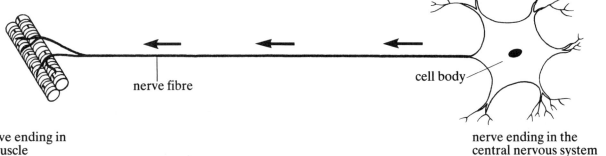

nerve fibre   cell body

nerve ending in
a muscle

nerve ending in the
central nervous system

Questions

1 How are responses co-ordinated in an animal?
2 Name two types of nerve cells.
3 What is an impulse?
4 Which type of nerve cell carries an impulse from a sense organ to the central nervous system?

Reflex actions

The fastest response to a stimulus is a **reflex action**. The response takes place before the person is conscious of what is happening. If you prick your finger on a pin you move your hand away quickly. This is a reflex action, which is **controlled by the spinal cord**.

stimulus — pain

response – the hand moves away

an impulse passes along a **sensory nerve fibre** to the spinal cord

an impulse passes along a **motor nerve fibre** to a muscle in the arm

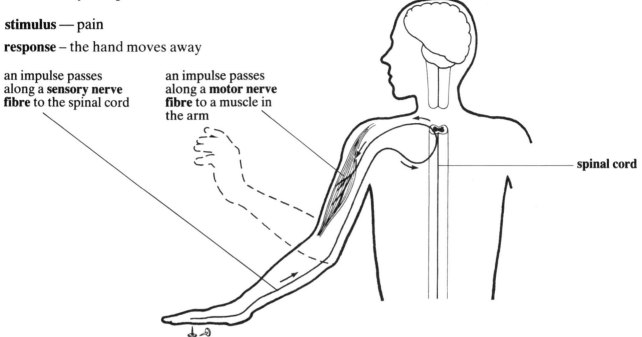

spinal cord

Section through the spinal cord

In the spinal cord there is a nerve cell in between the sensory nerve and the motor nerve.

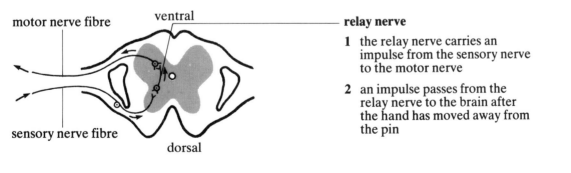

motor nerve fibre

ventral

sensory nerve fibre

dorsal

relay nerve

1 the relay nerve carries an impulse from the sensory nerve to the motor nerve

2 an impulse passes from the relay nerve to the brain after the hand has moved away from the pin

Questions

1 What is a reflex action?
2 In your own words, explain what happens when you put your hand on a hot saucepan.

The brain

Most responses are controlled by the brain. Different parts of the brain control different activities.

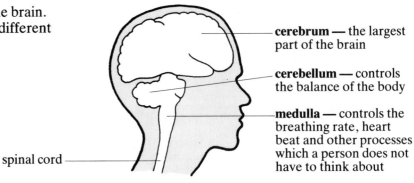

cerebrum — the largest part of the brain

cerebellum — controls the balance of the body

medulla — controls the breathing rate, heart beat and other processes which a person does not have to think about

spinal cord

The cerebrum

When a sense organ is stimulated an impulse passes along a sensory nerve to the cerebrum. The body's response to the stimulus is controlled by the cerebrum. The cerebrum also controls thinking, speech, learning and memory. The cerebrum is made of two parts called **cerebral hemispheres**.

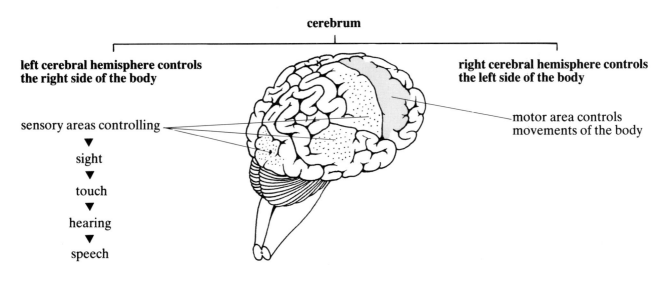

cerebrum

left cerebral hemisphere controls the right side of the body

right cerebral hemisphere controls the left side of the body

sensory areas controlling

▼

sight

▼

touch

▼

hearing

▼

speech

motor area controls movements of the body

Questions

1 What does the cerebellum control?
2 Which activities are controlled by the medulla?
3 Which part of the brain controls movements on the left side of the body?
4 Which activities are controlled by the motor area of the cerebrum?
5 Which activities are controlled by the sensory area of the cerebrum?

Co-ordination by hormones

Many processes are controlled by chemicals called hormones. Hormones are made in **endocrine glands**. The blood carries hormones from the endocrine glands to all parts of the body. Two hormones made by endocrine glands are **adrenalin** and **growth hormone**.

Adrenalin

The hormone **adrenalin** is made by the **adrenal glands**. Adrenalin is made when a person is frightened or angry.

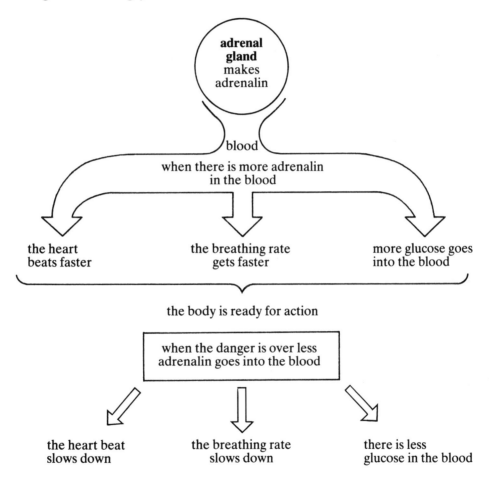

Questions

1 Where are hormones made?
2 Where is adrenalin made?
3 Explain what happens when there is a lot of adrenalin in the blood.

Growth hormone

Growth hormone is made by the **pituitary gland**. Growth hormone affects the size of the body.

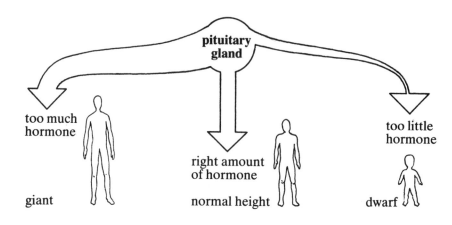

The endocrine glands

Endocrine gland	Hormone made by gland	What hormone does
pituitary	growth hormone	controls growth
Also called the master gland because it makes many hormones which control other glands in the body.		
thyroid	thyroxine	controls the chemical reactions in the cells of the body
adrenals	adrenalin	gets the body ready for action by speeding up the heart beat and breathing rate
pancreas	insulin	controls the amount of sugar (glucose) in the blood
ovaries (only in females)	female sex hormone	controls the female characteristics e.g. breasts
testes (only in males)	male sex hormone	controls the male characteristics e.g. deeper voice and hair on the body

---

Questions

1 Why is the pituitary gland called the master gland?
2 (a) Which hormone controls the amount of sugar in the blood?
  (b) Where is the hormone made?
3 Which gland makes the female sex hormone?
4 Which gland controls the male characteristics?
5 Where is thyroxine made?
6 Where is growth hormone made?
7 What happens if a child does not make enough growth hormone?

Summary

Responses in man are controlled or co-ordinated by the nervous system and the endocrine system.

The central nervous system is made up of the brain and the spinal cord. The central nervous system is linked to all parts of the body by nerve fibres. The nerve fibres carry impulses (messages) to and from all parts of the body. Responses co-ordinated by the nervous system take place quickly. The fastest response is a reflex action which is controlled by the spinal cord.

The endocrine system is made up of glands which make hormones. The blood carries hormones to all parts of the body. This makes the responses co-ordinated by the endocrine system slower than the responses co-ordinated by the nervous system.

Key words

central nervous system Brain and spinal cord

co-ordination Different parts of the body are controlled so that they work together

hormone A chemical which helps to co-ordinate different processes in the body e.g. growth and reproduction

impulse A message which passes along a nerve fibre

motor nerve fibre Carries impulses away from the central nervous system

reflex A response which happens quickly and without thought. e.g. pulling the hand away from a hot kettle

response A change in activity which has been caused by a stimulus

sensory nerve fibre Carries impulses from a sense organ to the central nervous system

stimulus Anything which causes a response e.g. pain

Questions

1 Reflex actions happen very quickly in response to stimuli. Match up the right response to each stimulus.

Stimulus

A a tap on the knee

B dust blowing into the eyes

C the smell of food

D going into a very dark room

Response

you blink

more saliva is made in the mouth

the pupil gets larger

the leg straightens

2 Look carefully at cell A and cell B.

   (a) Name each cell
   (b) Explain what each cell does
   (c) Name the parts labelled P, Q, R and S

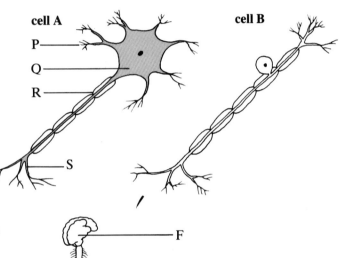

3 (a) Which part of the nervous system is shown in this diagram.
   (b) Name the parts labelled F and G.
   (c) What are the functions of the parts F and G.
   (d) If a person is injured at the part labelled X what affect would this have on parts of the body below the injury.

4 (a) Copy this outline of the body.
   (b) Show the position of four different endocrine glands.
   (c) Name each gland.
   (d) Name one hormone made by each gland.
   (e) How are hormones carried round the body?

17: Plant reproduction

A plant does not live for ever and therefore more plants must be produced to take the place of those which die. New plants are produced by **reproduction**.

Some plants reproduce by **asexual reproduction**. In asexual reproduction part of the parent plant separates to form a new plant. A new plant formed in this way will be exactly the same as its parent. This type of reproduction is also called **vegetative reproduction**.

Vegetative reproduction

The strawberry, crocus, potato, daffodil and Bryophyllum all reproduce by vegetative reproduction.

A strawberry runner

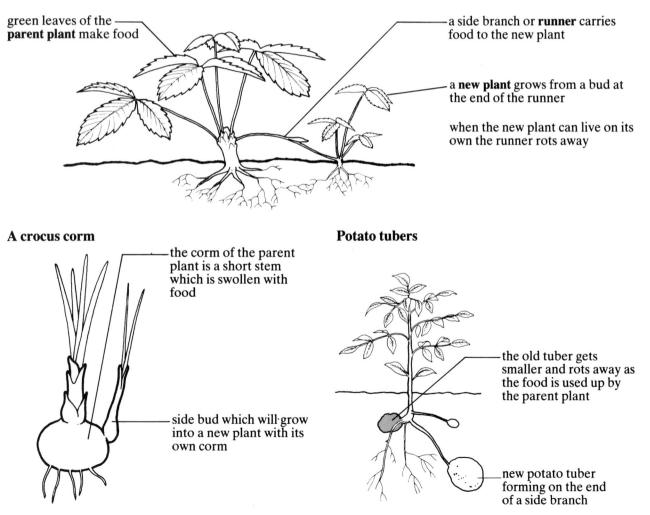

green leaves of the **parent plant** make food

a side branch or **runner** carries food to the new plant

a **new plant** grows from a bud at the end of the runner

when the new plant can live on its own the runner rots away

A crocus corm

the corm of the parent plant is a short stem which is swollen with food

side bud which will grow into a new plant with its own corm

Potato tubers

the old tuber gets smaller and rots away as the food is used up by the parent plant

new potato tuber forming on the end of a side branch

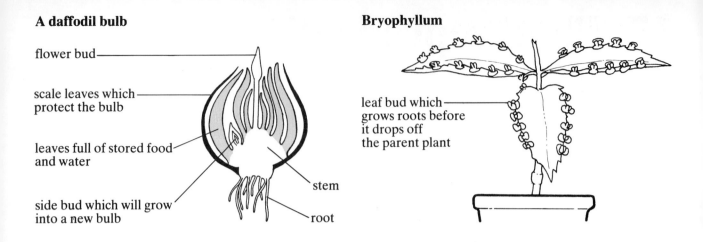

A daffodil bulb

flower bud

scale leaves which protect the bulb

leaves full of stored food and water

side bud which will grow into a new bulb

stem

root

Bryophyllum

leaf bud which grows roots before it drops off the parent plant

Questions

1 Why must plants and animals reproduce?
2 Name a plant which has a runner.
3 What does a runner do?
4 Name a plant which uses a corm for vegetative reproduction.
5 Name a plant which has a tuber.

Sexual reproduction in flowering plants

Flowering plants reproduce **sexually**. Most plants have male and female parts on the same flower.

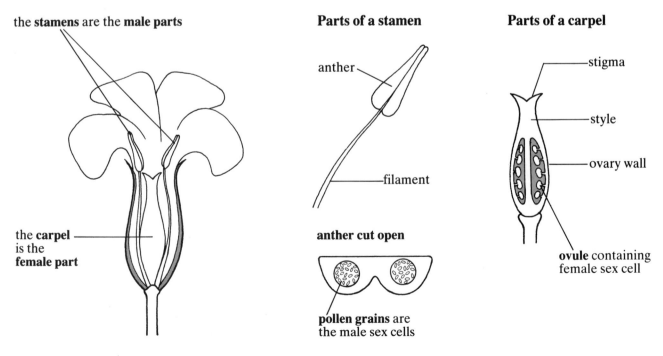

the **stamens** are the **male parts**

the **carpel** is the **female part**

Parts of a stamen

anther

filament

anther cut open

pollen grains are the male sex cells

Parts of a carpel

stigma

style

ovary wall

ovule containing female sex cell

The parts of a flower

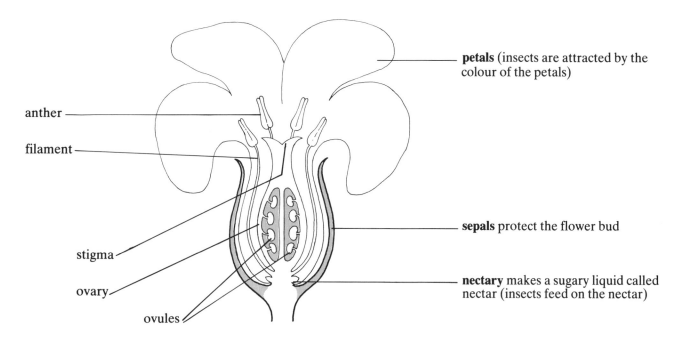

petals (insects are attracted by the colour of the petals)

anther

filament

stigma

ovary

ovules

sepals protect the flower bud

nectary makes a sugary liquid called nectar (insects feed on the nectar)

Questions

1 What are the male parts of a flower called?
2 What is a carpel?
3 Which part of the stamen has pollen grains in it?
4 Where is the female sex cell found?
5 Which part of a flower attracts insects?
6 What is nectar?

Pollination

Pollination takes place when **pollen from an anther** is carried **to a stigma**.

When pollen is carried to the stigma of the same flower it is called **self pollination**.

When pollen is carried to the stigma of another flower it is called **cross pollination**.

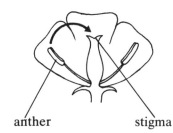

anther   stigma

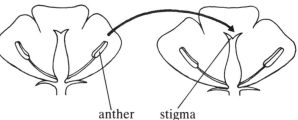

anther   stigma

Insect pollination

Pollen can be carried from one flower to another by insects. Wallflowers, buttercups and sweet peas are pollinated by insects.

pollen is carried by the bee

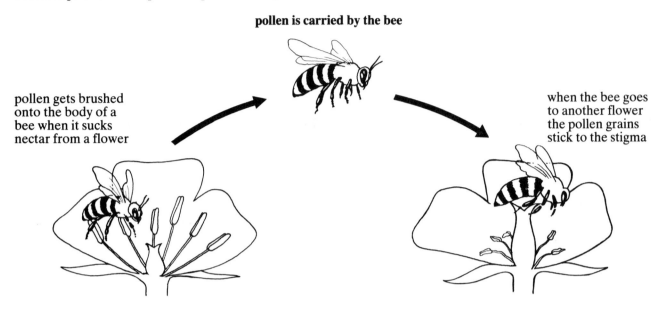

pollen gets brushed onto the body of a bee when it sucks nectar from a flower

when the bee goes to another flower the pollen grains stick to the stigma

Wind pollination

Pollen can be carried from one flower to another by the wind. Grasses, plantain and some trees are pollinated by the wind.

pollen is carried by the wind

large anthers on long filaments hang outside the flower. The pollen is blown away by the wind

pollen sticks to the feathery stigmas

150

The differences between insect and wind pollinated flowers

Insect pollinated flowers

These flowers have:

1 **large** brightly coloured **petals**

2 **scent**

3 **nectar**

4 **large pollen grains** with a rough surface to help them stick to the insect's body

5 **anthers and stigmas inside the flower** so that they touch the insect when it tries to get the nectar

Wind pollinated flowers

These flowers have:

1 very small petals

2 no scent

3 no nectar

4 a lot of small pollen grains with a smooth surface

5 large anthers and stigmas which hang outside the flower

Questions

1 Explain what happens when a flower is self pollinated.
2 How can an insect help to pollinate a flower?
3 Why do wind pollinated flowers have long feathery stigmas?
4 Give three differences between insect and wind pollinated flowers.
5 Name two flowers pollinated by wind.
6 Name three flowers pollinated by insects.

Fertilization of a flower

Sex cells are called **gametes**. The male gamete is in a pollen grain. The female gamete is in an ovule. A flower is fertilized when a male gamete joins up with a female gamete.

1 the pollen grain sticks to the stigma

2 the pollen grain bursts and a tube grows down the style

3 the tip of the pollen tube bursts and the male gamete joins up with the female gamete

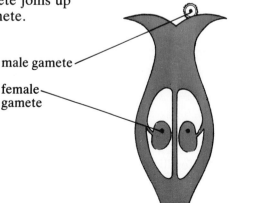

male gamete

female gamete

The life cycle of a pea

A pea flower cut in half

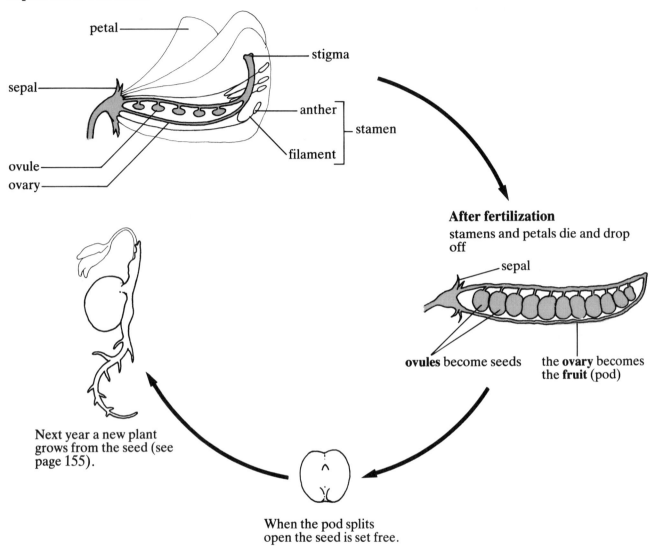

petal

stigma

sepal

anther

stamen

filament

ovule

ovary

After fertilization
stamens and petals die and drop off

sepal

ovules become seeds

the **ovary** becomes the **fruit** (pod)

When the pod splits open the seed is set free.

Next year a new plant grows from the seed (see page 155).

Questions

1 What is the male gamete of a flower?
2 What is the female gamete of a flower?
3 Explain how a flower is fertilized.
4 What happens to the ovary after a pea flower has been fertilized?
5 Which part of the flower becomes the seed after fertilization?
6 What happens when the ovary of a pea splits?

Seed dispersal

The fruits and seeds must be carried away from the parent plant. This is called **dispersal**. If the seeds are dispersed and carried away from the parent plant they will have enough space, light and water to grow next year.

Fruits and seeds can be dispersed by the plant itself, by the wind or by animals.

Seeds dispersed by the plant itself

When the ovary wall (pod) dries out it splits open and the seeds are scattered away from the plant.

pea

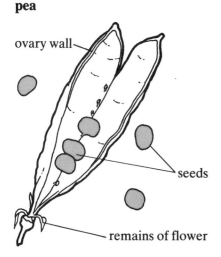

ovary wall

seeds

remains of flower

wallflower

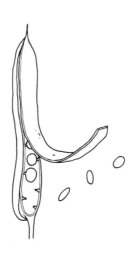

pansy

remains of style

remains of flower

seeds

Seeds dispersed by the wind

dandelion

parachute of hairs helps to keep the fruit in the air

fruit

sycamore

wing made by the ovary wall helps to carry the fruit away from the parent plant

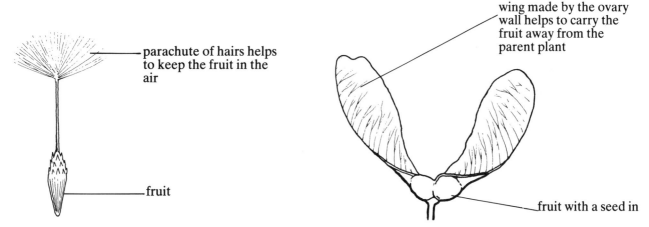

fruit with a seed in

Seeds dispersed by animals

Birds and people eat the soft parts of an apple and leave the seeds behind.

apple

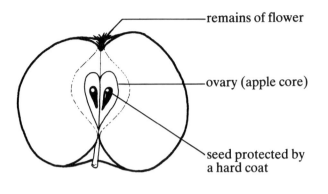

- remains of flower
- ovary (apple core)
- seed protected by a hard coat

burdock

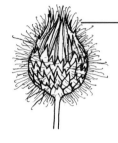

hooks which catch onto an animal's fur

acorn

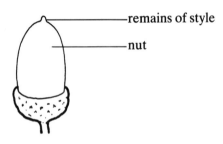

- remains of style
- nut

some acorns buried by squirrels grow into oak trees

strawberry

strawberry seeds can go through the digestive system of an animal without being harmed

---

Questions

1 Explain why seeds are dispersed.
2 How does a pea plant disperse its seeds?
3 **(a)** Name two plants which use wind dispersal.
   (b) Explain how their fruits are suited to wind dispersal.
4 **(a)** Explain how animals can help to disperse seeds.
   (b) Name three plants which are dispersed by animals.

Germination

Most seeds do not start to grow as soon as they are dispersed. They stay **dormant** during the autumn and winter. In the spring they start to grow or **germinate**. A seed will germinate if it gets enough **water**, **air** and **warmth**.

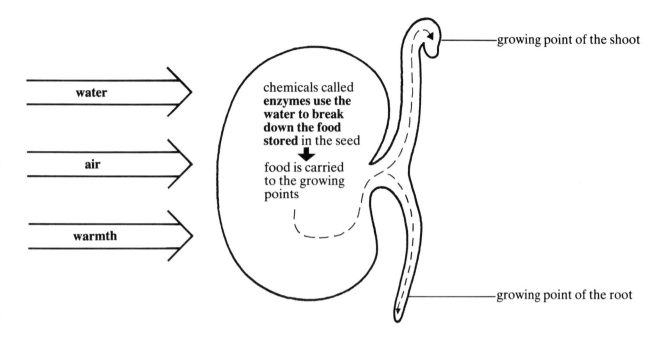

The French bean seed

Food is stored in the seed leaves or **cotyledons** of a seed. A French bean seed has two cotyledons it is therefore called a **dicotyledon**.

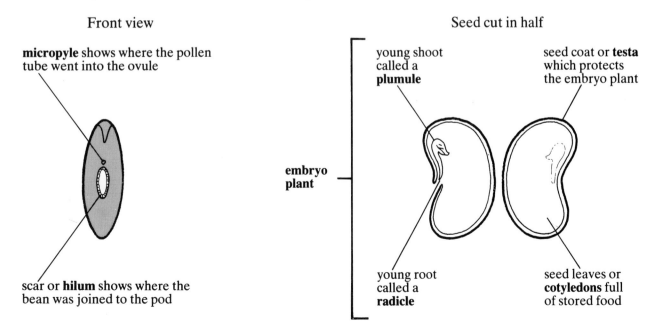

155

Germination of a French bean

When a French bean germinates the **cotyledons come above the ground**. This is called **epigeal germination**.

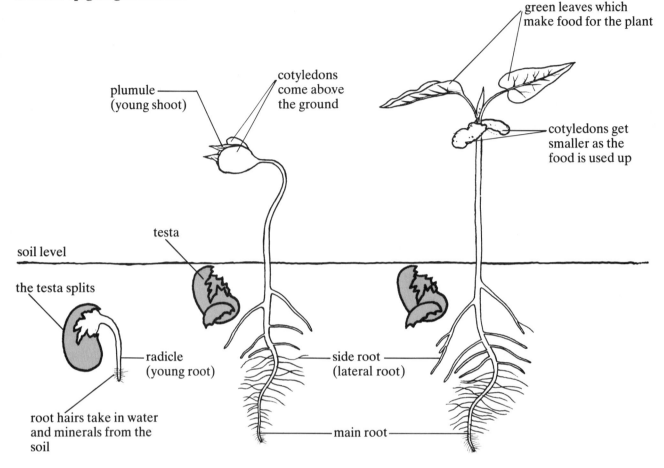

green leaves which make food for the plant

plumule (young shoot)

cotyledons come above the ground

cotyledons get smaller as the food is used up

testa

soil level

the testa splits

radicle (young root)

side root (lateral root)

root hairs take in water and minerals from the soil

main root

The maize grain

A maize grain has only one cotyledon therefore it is called a **monocotyledon**. The food needed for growth is stored in the **endosperm**.

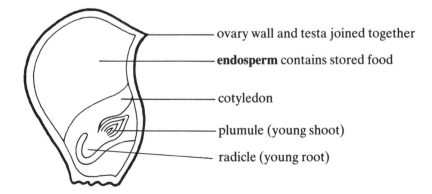

ovary wall and testa joined together

endosperm contains stored food

cotyledon

plumule (young shoot)

radicle (young root)

Germination of a maize grain

When a maize grain germinates the **cotyledon stays below the ground**. This is called **hypogeal germination**.

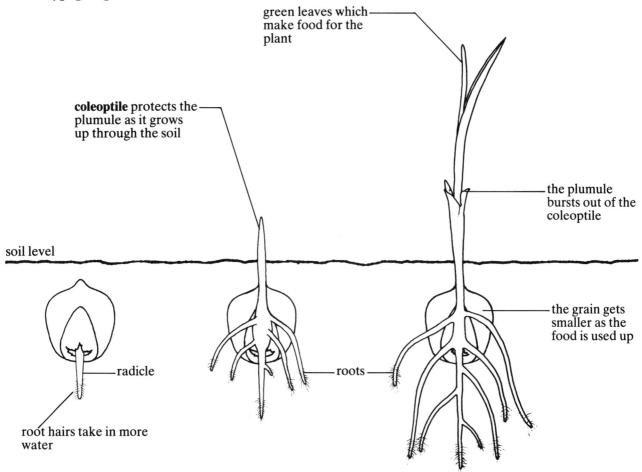

green leaves which make food for the plant

coleoptile protects the plumule as it grows up through the soil

the plumule bursts out of the coleoptile

soil level

the grain gets smaller as the food is used up

radicle

roots

root hairs take in more water

Questions

1 What is **(a)** a radicle and **(b)** a plumule?
2 What does the testa on a French bean seed do?
3 Where is food stored in a French bean seed?
4 Why must a seed have water before it can germinate?
5 Which part of a seed grows first **(a)** the radicle or **(b)** the plumule?
6 Where is a food stored in a maize grain?
7 Explain why **(a)** a French bean is a dicotyledon and **(b)** maize is a monocotyledon.
8 What is the main difference between epigeal and hypogeal germination?

Summary

Some flowering plants can reproduce asexually by producing bulbs, corms, runners or tubers which will grow into new plants.

Flowers are the reproductive organs of a plant. In most cases the stamens (male parts) and carpels (female parts) are on the same flower. A flower is pollinated when pollen (the male sex cell) is carried onto the part of the carpel called the stigma. Large brightly coloured flowers with nectar are pollinated by insects. Small dull flowers without nectar are wind pollinated. After pollination the flower is fertilized. The nucleus from a pollen grain joins up with the nucleus in the ovule.

After fertilization the ovule develops into a seed and the ovary wall becomes a fruit. The seed is dispersed by the wind, animals or the plant itself. After a dormant period the seed germinates into a new plant.

Key words

anther	Part of the stamen which makes pollen grains
carpels	Female parts of a flower
dicotyledon	A plant with two cotyledons in its seed e.g. bean, pea
epigeal germination	The cotyledons come above the ground
fertilization	Nucleus from a pollen grain joining up with the nucleus in an ovule
hypogeal germination	The cotyledons stay below the ground
monocotyledon	A plant with one cotyledon in its seed e.g. maize
ovary	Part of a carpel where ovules are made. It develops into a fruit after fertilization
ovule	Part of a carpel containing the female sex cell. It develops into a seed after fertilization
pollen grain	The male sex cell of a plant
pollination	Pollen grains are carried from an anther to a stigma
stamens	Male parts of a flower
stigma	Part of the carpel where pollen lands
vegetative reproduction	A kind of asexual reproduction e.g. when a new plant grows from a bulb, corm, runner or tuber

Questions

1 Give two example of fruits that disperse their seeds by splitting a pod.

2 These diagrams show two different types of seed dispersal.

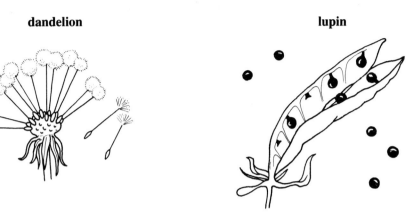

dandelion   **lupin**

Explain why the seeds of the dandelion will be dispersed further away from the parent plant than those of the lupin.

3 Explain how some plants reproduce asexually.

4 **(a)** Make a large labelled diagram of an insect pollinated flower.
(b) Give three differences between insect and wind pollinated flowers.

5 Copy and complete this paragraph with the following words:–

anther   seed   pollen   fruit

style   stigma   fertilized   gamete

A flower is pollinated when............ is carried from the............ of a stamen to the............ of a carpel. A pollen tube grows down the............ until it reaches an ovule. The tip of the tube bursts and the male............ joins up with the female gamete in the ovule. When this happens the ovule is............ The ovule then develops into a............ and the ovary becomes the............

6 Make a clear labelled drawing of a seed you have studied to show as many parts as possible.

7 Look at these diagrams showing how a pea seed germinates. Describe what is happening at each stage. Include the following words in your answer:

testa   radicle   plumule   cotyledon   root hairs

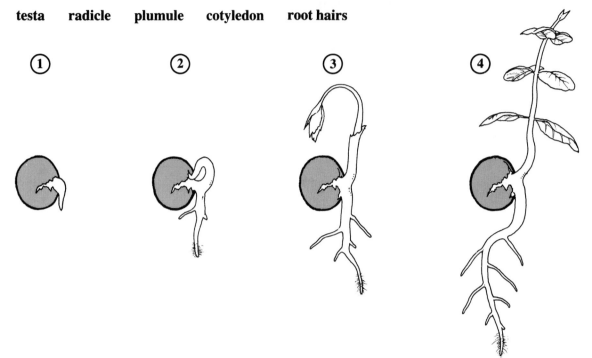

8 Explain why some of the cress seeds used for an experiment did not germinate (grow). Tubes A, B and C had been kept in a warm room for ten days. Tube D had been kept in a refrigerator for ten days.

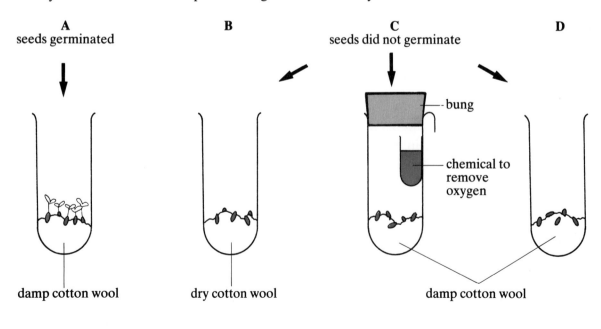

18: Animal reproduction

An animal does not live for ever. Many are eaten or killed by disease, others die from old age. More animals must be produced to take the place of those which die. More animals are produced by **reproduction**. Most animals reproduce by **sexual reproduction**.

The sex cells are called **gametes**. The male gamete is called a **sperm**. The female gamete is called an egg or **ovum**. Fertilization takes place when the nucleus from a male gamete joins up with the nucleus from a female gamete.

Reproduction in fish – the herring

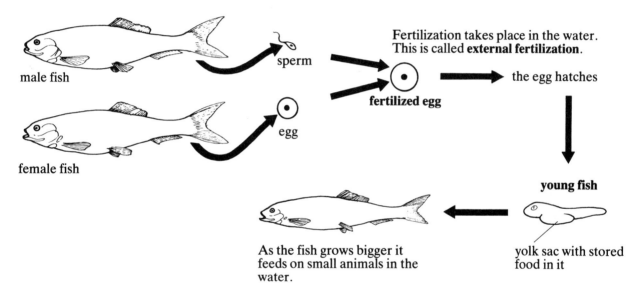

male fish

sperm

female fish

egg

Fertilization takes place in the water. This is called **external fertilization**.

fertilized egg

the egg hatches

young fish

yolk sac with stored food in it

As the fish grows bigger it feeds on small animals in the water.

A female herring lays many eggs because:

1 some eggs will not be fertilized by sperm
2 some eggs will be eaten by other fish
3 some fish will be eaten or die before they are old enough to reproduce

Questions

1 What is a sperm?
2 What is another name for an egg?
3 How is an egg fertilized?

4 Where does a newly hatched fish get its food from?
5 Explain why a female herring must lay many eggs.

Reproduction in amphibia – the frog

Frogs can live on land, but they must go back to the water to lay their eggs. An egg hatches into a tadpole or **larva**. The larva looks very different from the adult frog. The change from a larva to an adult is called **metamorphosis**.

Frogs start breeding in the spring. A male frog climbs onto the back of a female frog. As the female lays her eggs the male releases his sperms into the water. Fertilization takes place in the water.

Metamorphosis of a frog

The male frog has large pads on his thumbs. This helps him to hold on to the female.

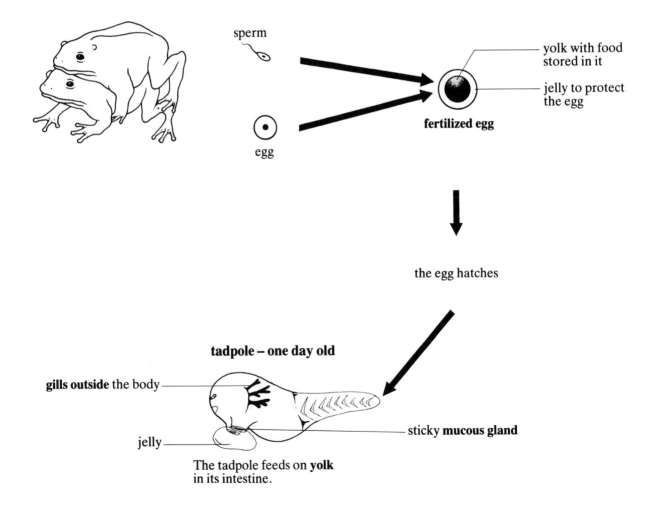

sperm

yolk with food stored in it

jelly to protect the egg

egg

fertilized egg

the egg hatches

tadpole – one day old

gills outside the body

jelly

sticky **mucous gland**

The tadpole feeds on **yolk** in its intestine.

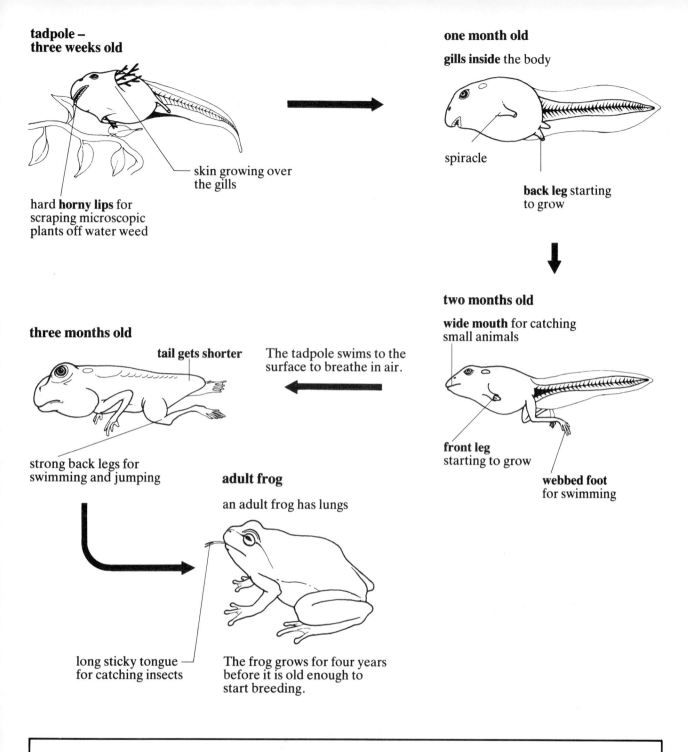

tadpole – three weeks old

skin growing over the gills

hard **horny lips** for scraping microscopic plants off water weed

one month old

gills inside the body

spiracle

back leg starting to grow

two months old

wide mouth for catching small animals

front leg starting to grow

webbed foot for swimming

three months old

tail gets shorter

The tadpole swims to the surface to breathe in air.

strong back legs for swimming and jumping

adult frog

an adult frog has lungs

long sticky tongue for catching insects

The frog grows for four years before it is old enough to start breeding.

Questions

1 What does a tadpole feed on when it is:
   (a) one day old
   (b) three weeks old
   (c) two months old?

2 How old is a tadpole when it:
   (a) has gills outside the body
   (b) has gills inside the body
   (c) swims to the surface to breathe in air?

Reproduction in insects

Some insects have young which **look very different** from the adult. The young insect gradually **changes** into an adult. This type of development is called **metamorphosis**. Flies, moths and butterflies develop in this way.

Some insects have young which **look similar** to the adult. The young insect gradually **grows larger** until it becomes an adult. This type of development is called **incomplete metamorphosis**. Locusts, cockroaches and dragonflies develop in this way.

Metamorphosis – the cabbage white butterfly

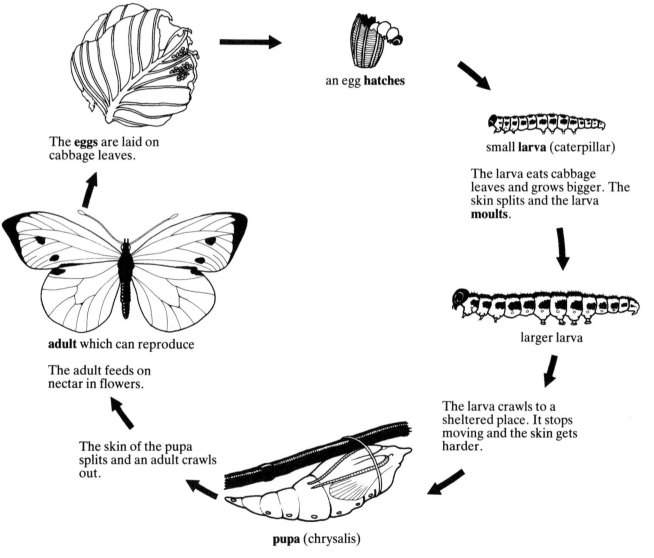

The **eggs** are laid on cabbage leaves.

an egg **hatches**

small **larva** (caterpillar)

The larva eats cabbage leaves and grows bigger. The skin splits and the larva **moults**.

larger larva

The larva crawls to a sheltered place. It stops moving and the skin gets harder.

adult which can reproduce

The adult feeds on nectar in flowers.

The skin of the pupa splits and an adult crawls out.

pupa (chrysalis)

The larval tissues are broken down. The adult tissues are built up.

Incomplete metamorphosis – the locust

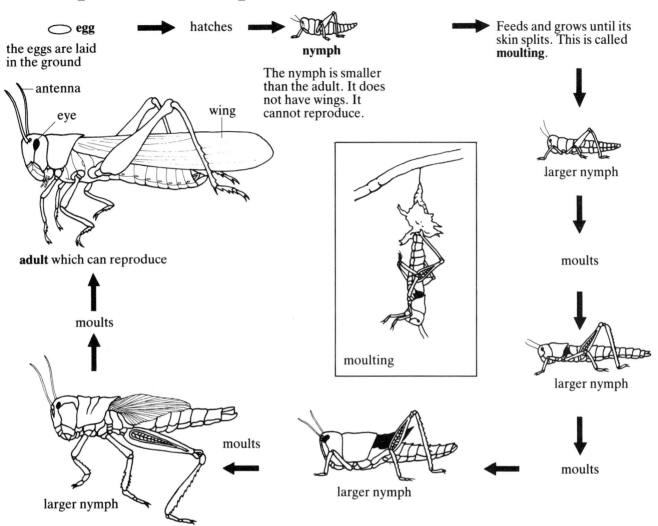

egg
the eggs are laid
in the ground

hatches

nymph

The nymph is smaller
than the adult. It does
not have wings. It
cannot reproduce.

Feeds and grows until its
skin splits. This is called
moulting.

antenna

eye

wing

adult which can reproduce

moults

larger nymph

moulting

larger nymph

moults

larger nymph

moults

larger nymph

moults

moults

Questions

1 Copy and complete these sentences:

If an insect has young which look very
different from the adult they develop
by............ Flies and............ develop in
this way.

If an insect has young which look similar
to the adult they develop by.............
Dragonflies and............ develop in this
way.

2 Where does a cabbage white butterfly lay
its eggs?
3 What does the larva of a cabbage white
butterfly feed on?
4 What does the adult cabbage white
butterfly feed on?
5 Give another name for the pupa of a
butterfly.
6 How does the nymph of a locust differ
from the adult?
7 What is moulting?

Reproduction in birds – the chaffinch

1 Courtship

The **male bird attracts the female** by:

(a) singing

(b) showing off brightly coloured feathers

2 Mating

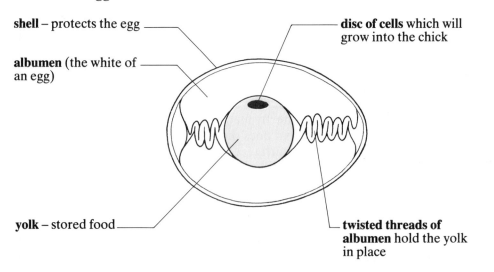

male bird

female bird

Sperms are put into the female when the male reproductive opening is pressed against the female reproductive opening. Fertilization takes place inside the female. This is called **internal fertilization**.

3 Fertilized egg

shell – protects the egg

disc of cells which will grow into the chick

albumen (the white of an egg)

yolk – stored food

twisted threads of albumen hold the yolk in place

166

4 Incubation

The parents sit on the eggs to keep them warm.

5 Development of the chick

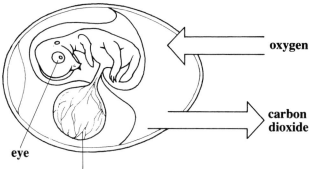

oxygen

carbon dioxide

eye

blood vessels which carry the food from the yolk into the chick

6 Hatching

The chick uses its **egg tooth** to break the shell.

7 Feeding

The young bird is **fed by the parents** until it is old enough to look for its own food.

Questions

1 How does a male chaffinch attract a female chaffinch?
2 Does fertilization of a chaffinch egg take place inside or outside the body?
3 Which part of an egg has food stored in it?
4 Explain how birds look after their young.

Reproduction in mammals – man

When a boy is 11-15 years old his **testes** start to make:

1 sperms
2 male sex hormone called **testosterone**

which makes **(a)** under-arm and pubic hair grow
              (b) more hair grow on the face and chest
              (c) the voice deeper

The male sex organs

Front view

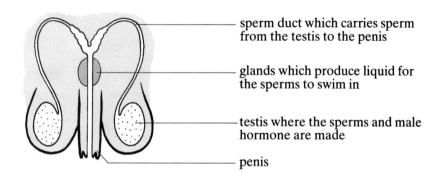

sperm duct which carries sperm from the testis to the penis

glands which produce liquid for the sperms to swim in

testis where the sperms and male hormone are made

penis

Side view

The arrows show how the sperms get from the testis to the penis.

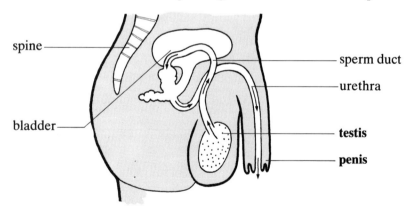

spine

bladder

sperm duct

urethra

testis

penis

Questions

1 How old is a boy when he starts to make sperms?
2 Where are sperms made?
3 Give two changes which will begin in a boy when the male sex
   hormone starts to go round his body.

The female sex organs

When a girl is 8-15 years old her **ovaries**:

1 start to release **ova** (eggs)
2 start to make **female sex hormone** called **oestrogen**

which makes **(a)** the breasts grow larger
                (b) under-arm and pubic hair grow

Side view of female sex organs

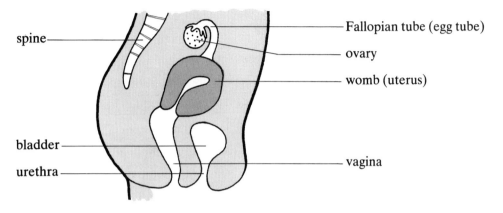

Front view of female sex organs

The arrows show how an ovum gets from the ovary to the womb.

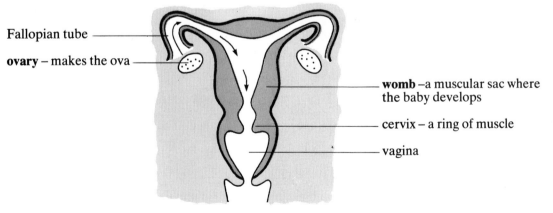

Questions

1 How old is a girl when she starts to release ova?

2 Where are ova made?

3 What is another name for the womb?

4 What is the cervix?

5 Give two changes which will begin in a girl when female hormones start to go round her body.

The menstrual cycle

As soon as the ovaries start to release ova some changes take place inside the womb. These changes prepare the womb for a fertilized ovum.

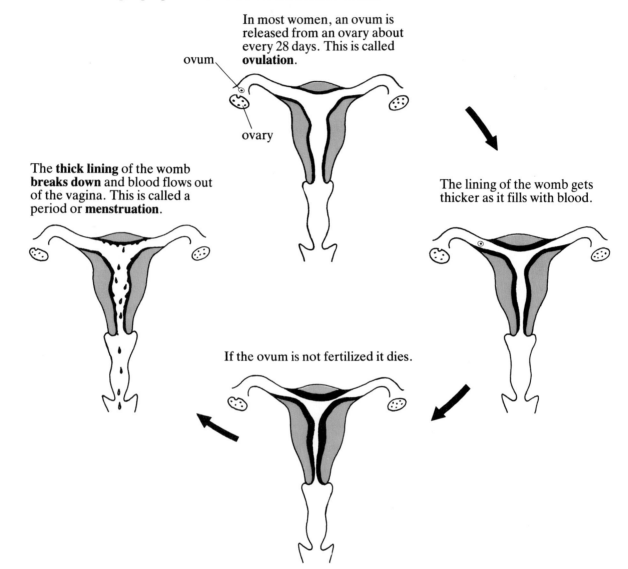

In most women, an ovum is released from an ovary about every 28 days. This is called **ovulation**.

ovum

ovary

The lining of the womb gets thicker as it fills with blood.

The **thick lining** of the womb **breaks down** and blood flows out of the vagina. This is called a period or **menstruation**.

If the ovum is not fertilized it dies.

This regular 28 day cycle is called the **menstrual cycle**. After an ovum has been released the woman is **fertile** and she may become **pregnant** if she has sexual intercourse.

Questions

1 How often is an ovum released from an ovary?
2 What happens to the lining of womb after an ovum has been released?
3 What happens to the lining of the womb if the ovum is not fertilized?
4 What is menstruation?

Fertilization

Fertilization takes place in the Fallopian tube.

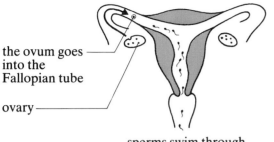

the ovum goes into the Fallopian tube

ovary

sperms swim through the womb and up into the Fallopian tube

1 Sperms are attracted to the ovum.

2 Only one sperm breaks the surface of the ovum.

3 The sperm nucleus joins up with the ovum nucleus.

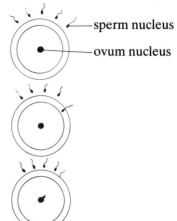

sperm nucleus

ovum nucleus

Development of a fertilized ovum

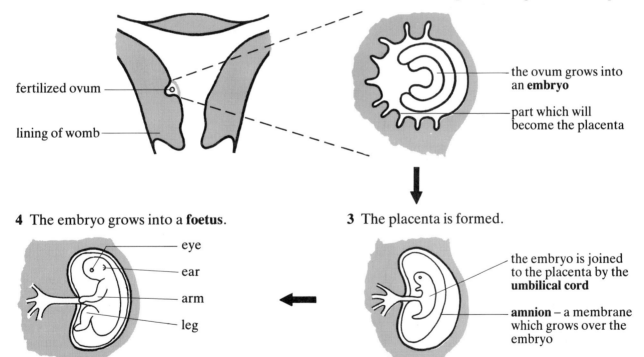

1 The fertilized ovum goes to the womb.

fertilized ovum

lining of womb

2 The ovum buries itself in the thick lining of the womb. The **placenta** begins to develop.

the ovum grows into an **embryo**

part which will become the placenta

3 The placenta is formed.

the embryo is joined to the placenta by the **umbilical cord**

amnion – a membrane which grows over the embryo

4 The embryo grows into a **foetus**.

eye

ear

arm

leg

Questions

1 Where does fertilization take place?
2 Explain what happens when an ovum is fertilized.

3 How many sperms fertilize an ovum?
4 Explain what happens after an ovum has been fertilized.

The placenta

The mother's blood does not mix with the embryo's blood. Substances are exchanged through the placenta. Food and oxygen from the mother's blood pass through the placenta to the embryo. Carbon dioxide and other waste substances from the embryo pass through the placenta to the mother's blood. The placenta also stops harmful substances passing into the embryo.

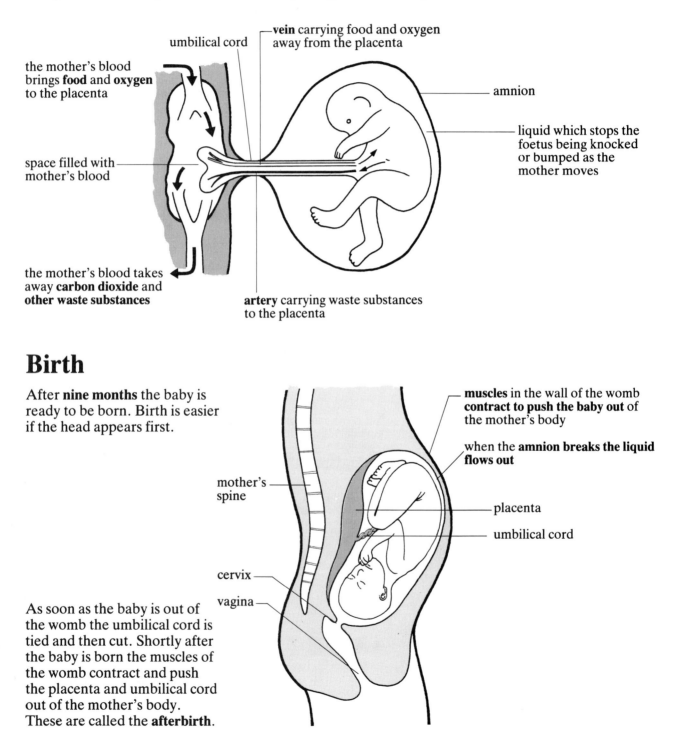

umbilical cord

vein carrying food and oxygen away from the placenta

the mother's blood brings **food** and **oxygen** to the placenta

amnion

liquid which stops the foetus being knocked or bumped as the mother moves

space filled with mother's blood

the mother's blood takes away **carbon dioxide** and **other waste substances**

artery carrying waste substances to the placenta

Birth

After **nine months** the baby is ready to be born. Birth is easier if the head appears first.

muscles in the wall of the womb **contract to push the baby out** of the mother's body

when the **amnion breaks the liquid flows out**

mother's spine

placenta

umbilical cord

cervix

vagina

As soon as the baby is out of the womb the umbilical cord is tied and then cut. Shortly after the baby is born the muscles of the womb contract and push the placenta and umbilical cord out of the mother's body. These are called the **afterbirth**.

Summary

Most animals reproduce by sexual reproduction. Sperms are the male gametes. Ova or eggs are the female gametes. An ovum is fertilized when the nucleus from a male gamete joins up with the nucleus from a female gamete.

In fish and amphibians fertilization takes place outside the body. These animals produce a lot of eggs because many will not be fertilized. Birds and mammals do not produce as many eggs because fertilization takes place inside the female. Birds and mammals also look after their young until they are old enough to feed themselves.

In mammals like man the fertilized ovum buries itself in the lining of the womb and develops inside the mother. Food and oxygen from the mother's blood pass through the placenta to the developing embryo. The embryo's waste substances pass through the placenta and are carried away by the mother's blood.

If an insect develops from a larva which looks very different from an adult it is called metamorphosis. If an insect develops from a nymph which looks similar to an adult it is called incomplete metamorphosis.

Key words

fertilization Male and female sex cells joining together during sexual reproduction

larva A stage in the life cycle of some animals. A larva looks very different from an adult e.g. tadpole of a frog. caterpillar of a butterfly

metamorphosis When a larva changes into an adult

nymph A stage in the life cycle of some insects e.g. locust. A nymph looks similar to an adult but does not have wings and cannot reproduce

ovary Where female sex cells are made

ovum A female sex cell or egg (many ova)

placenta An embryo gets food and oxygen from the mother through the placenta. Waste substances from the embryo pass through the placenta and are carried away by the mother's blood

pupa Chrysalis stage in the life cycle of a butterfly

sperm A male sex cell

testis Where male sex cells are made

Questions

1 (a) Draw a large labelled diagram of a woman's reproductive organs.
   (b) Label the ovary, the womb and the Fallopian tube.
   (c) On your diagram use arrows to show how the sperms reach an ovum.
   (d) Put an X on your diagram to show where fertilization takes place.
   (e) Describe how a sperm fertilizes an ovum.
   (f) Women who do not want any more children sometimes have their Fallopian tubes cut and tied. Explain how this stops an ovum being fertilized.

2 (a) Draw a diagram of the male reproductive system.
   (b) Label (1) a testis (2) the urethra (3) the penis (4) the bladder
   (c) Where are sperms made?
   (d) On your diagram use arrows to show how sperms reach the penis.

3 Copy and complete this paragraph with the following words:

yolk   animals   jaws   wider   sperms   eggs   larva

male   fertilized   metamorphosis   microscopic plants

In spring the .... frog climbs on the female's back. The female lays about one thousand............ The eggs are............ by............ from the male. A tadpole or............ hatches after about ten days. At first the tadpole feeds on............ in its intestine. Later it uses horny............ to scrape............ off water weeds. When the tadpole is about two months old its mouth grows............ and it feeds on small............ in the water. When it is about three months old the larva changes into an adult. This series of changes is called............

4 (a) Describe the position of a baby when it is ready to be born.
(b) What happens when the amnion breaks?
(c) How is a baby pushed out of the womb?

5 (a) Explain the menstrual cycle.
(b) What is ovulation?
(c) What is menstruation?

6 (a) Which part of a bird's egg protects the developing embryo?
(b) How is an embryo in a female rabbit protected from knocks and bumps when the mother moves?

7 A developing embryo needs a supply of food. Explain how the following embryos get their food (a) a frog (b) a bird (c) a rabbit.

8 In some animals fertilization takes place outside the body (external fertilization). In others fertilization takes place inside the body (internal fertilization)
(a) Name two animals which use external fertilization.
(b) Name two animals which use internal fertilization.
(c) Explain why animals using external fertilization lay more eggs than those using internal fertilization.

9 What is meant by

(a) metamorphosis?
(b) incomplete metamorphosis?

10 Give one difference between a larva and a nymph.

11 With the help of labelled diagrams describe the stages in the life cycle of an insect you have studied.

19: Heredity

Some characteristics make a plant or an animal look like its parents. Other characteristics make a plant or an animal look different from its parents. These characteristics are **inherited** from the parents.

Eye colour is an inherited characteristic. The colour of the eyes makes some children in a family look like their parents. The colour of the eyes makes other children in a family look different from their parents.

father with **brown eyes**   mother with **brown eyes**

some children have **brown eyes** which makes them look like their parents

some children have **blue eyes** which makes them look different from their parents

Cell division

Before you can understand how characteristics like eye colour are inherited you need to know what happens when cells divide. There are long thread-like structures called **chromosomes** in the nucleus of a cell. The chromosomes are always in pairs. The next diagram explains what happens to one pair of chromosomes when a cell divides.

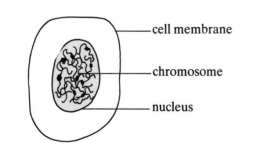

- cell membrane
- chromosome
- nucleus

two chromosomes in a cell

each chromosome splits into two parts

one half of each chromosome goes to each end of the cell

two new cells with two chromosomes in each

This type of cell division is called **mitosis**.

How sex cells are made

The male sex cells or sperms are made in the testes. The female sex cells or ova are made in the ovaries. Sperms are called **male gametes**, and ova are called **female gametes**. The cells which make the male and female gametes divide in a different way to other body cells.

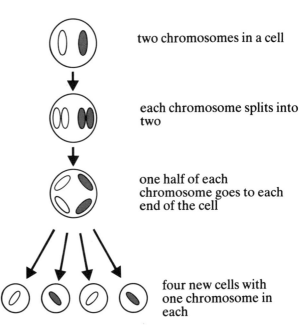

two chromosomes in a cell

each chromosome splits into two

one half of each chromosome goes to each end of the cell

four new cells with one chromosome in each

This type of cell division is called **meiosis**. When cells divide by meiosis the **number of chromosomes is halved**.

Fertilization

When **fertilization** takes place the **number of chromosomes is doubled**. This means that the fertilized ovum has the same number of chromosomes as all the other body cells.

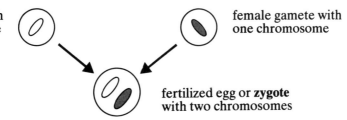

male gamete with one chromosome

female gamete with one chromosome

fertilized egg or **zygote** with two chromosomes

Questions

1 In which part of a cell are the chromosomes found?
2 (a) What are the male gametes called?
  (b) Where are they made?
3 (a) What are the female gametes called?
  (b) Where are they made?
4 Is the number of chromosomes halved when cells divide by (a) mitosis or (b) by meiosis?
5 Which type of cells divide by meiosis?

The inheritance of eye colour

Chromosomes are made up of many **genes**. The genes control the development of characteristics which we inherit from our parents. Genes are always in pairs. A pair of genes may be the same or they may be different.

A person with the genes **BB** will have **brown** eyes.
A person with the genes **bb** will have **blue** eyes.
A person with the genes **Bb** will have **brown** eyes.

In a person with the genes Bb only the effect of gene B shows. The effect of gene B **dominates** the effect of gene b. Gene B is therefore called the **dominant gene**. Gene b is called the **recessive gene**.

The next diagrams explain how the genes which control eye colour are inherited.

B is the gene for brown eyes.   b is the gene for blue eyes.

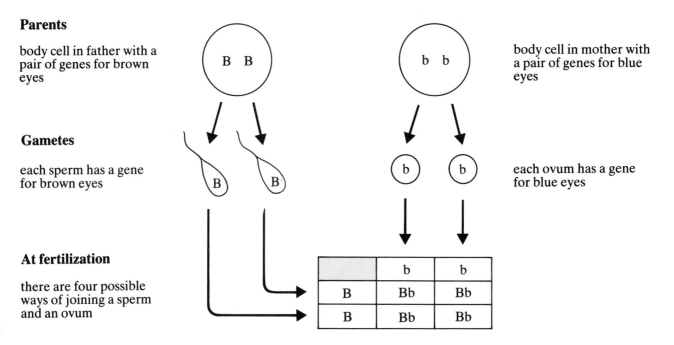

Parents

body cell in father with a pair of genes for brown eyes

body cell in mother with a pair of genes for blue eyes

Gametes

each sperm has a gene for brown eyes

each ovum has a gene for blue eyes

At fertilization

there are four possible ways of joining a sperm and an ovum

	b	b
B	Bb	Bb
B	Bb	Bb

All the children inherited the genes Bb. All the children have brown eyes.

Questions

1 Which part of a chromosome controls inherited characteristics?
2 Should you use a capital letter or a small letter for the dominant characteristic?

3 If B is the gene for brown eyes and b is the gene for blue eyes what colour eyes would the following people have:
(a) BB (b) Bb and **(c) bb?**

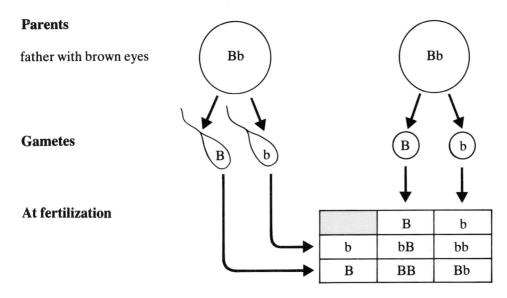

Parents

father with brown eyes — **Bb** ... **Bb** — mother with brown eyes

Gametes

At fertilization

	B	b
b	bB	bb
B	BB	Bb

a child who inherits the genes BB will have brown eyes
a child who inherits the genes Bb will have brown eyes
a child who inherits the genes bb will have blue eyes

In this case it is three times more likely that a child will have brown eyes.

Sex chromosomes

Every cell in the human body has 46 chromosomes. There are 22 pairs of chromosomes. The two other chromosomes do not always look alike. They are called the sex chromosomes. Female cells have two sex chromosomes which are alike (XX). Male cells have two chromosomes which are not alike (XY).

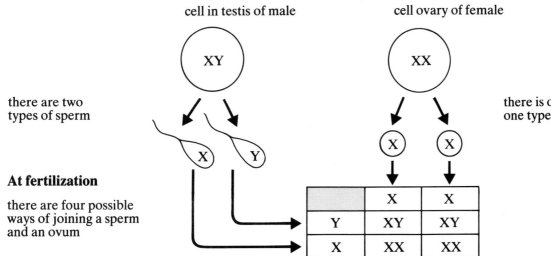

cell in testis of male ... cell ovary of female

there are two types of sperm

there is only one type of ovum

At fertilization

there are four possible ways of joining a sperm and an ovum

	X	X
Y	XY	XY
X	XX	XX

The sex of the baby depends on which type of sperm fertilizes the ovum. The chances of the baby being a boy (XY) or a girl (XX) are equal.

179

The work of Gregor Mendel

Just over one hundred years ago a monk called Gregor Mendel tried to discover how characteristics are inherited. For one of his experiments Mendel used pea plants which were either tall or dwarf (short).

Tall plants

1 Flowers on tall plants were pollinated with pollen from tall plants. Flowers on dwarf plants were pollinated with pollen from dwarf plants.

Dwarf plants

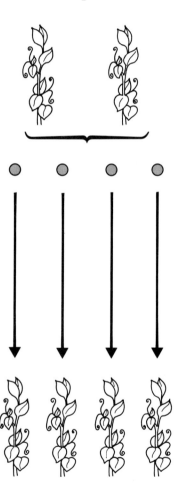

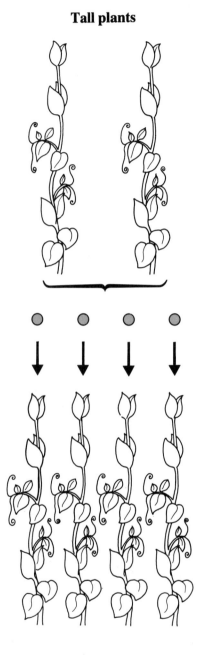

2 The seeds were collected.

3 The seeds were planted.

4 If all the new plants looked the same as the parent plants they were **pure breeding** (true breeding).

Mendel's breeding experiments

Mendel used the pure breeding plants for his next experiment. He pollinated the flowers on a pure breeding tall plant with pollen from a pure breeding dwarf plant. The plants grown from these seeds were called the **F₁ generation**. Tallness was the dominant characteristic in the F₁ generation. The F₁ generation tall plants were called **hybrids** because they inherited different characteristics from each parent.

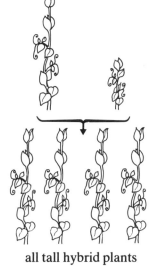

pure breeding tall plant

pure breeding dwarf plant

F₁ generation

all tall hybrid plants

The F₂ generation

Mendel pollinated the flowers on one hybrid plant with pollen from another hybrid plant. The new plants grown from these seeds were called the **F₂ generation**. Three quarters of the plants were tall and one quarter of the plants were dwarf. The dwarfness reappeared in the F₂ generation. Dwarfness was the recessive characteristic.

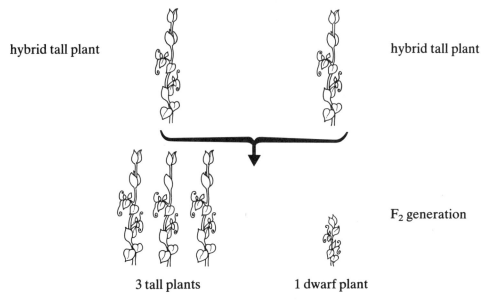

hybrid tall plant

hybrid tall plant

F₂ generation

3 tall plants

1 dwarf plant

The inheritance of stem length in pea plants

Mendel's experiments are explained in the following diagram.

A pure breeding tall plant will have the genes TT – it will look tall.

A pure breeding dwarf plant will have the genes tt – it will look dwarf.

A hybrid tall plant will have the genes Tt – it will look tall.

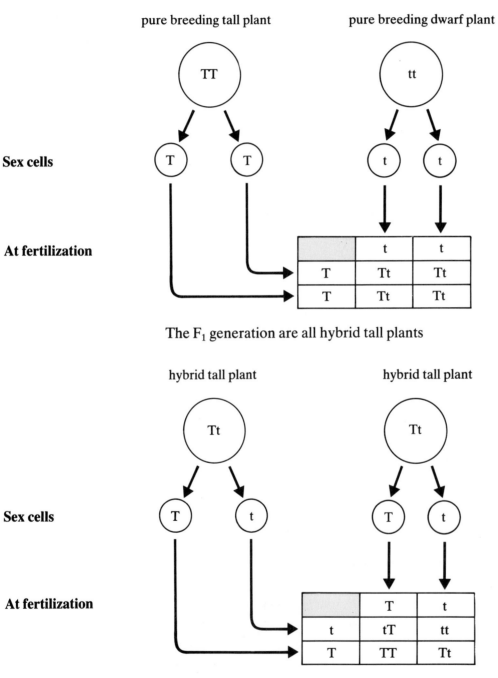

pure breeding tall plant   pure breeding dwarf plant

Sex cells

At fertilization

	t	t
T	Tt	Tt
T	Tt	Tt

The F_1 generation are all hybrid tall plants

hybrid tall plant   hybrid tall plant

Sex cells

At fertilization

	T	t
t	tT	tt
T	TT	Tt

In the F_2 generation it is three times more likely that the plant will be tall.

Questions

1 What would a plant with the genes tt look like?
2 Which genes would be found in a pure breeding tall plant?
3 What would the parents of a hybrid tall plant be like?
4 Did all the tall plants in the F_2 generation in Mendel's experiment have the same genes? Explain your answer.

Summary

The colour of your eyes and hair as well as the shape of your nose and ears are some of the characteristics you inherited from your parents.

Gregor Mendel was the first person to explain how hereditary characteristics pass from one generation to another. Other scientists have shown that each cell contains thread-like structures called chromosomes. Each chromosome carries many genes which control hereditary characteristics.

Sex cells (pollen, sperm, egg or ovum) are made when a parent cell divides by meiosis. In this type of cell division the pairs of chromosomes separate so that each sex cell has only half as many chromosomes as the parent cell. At fertilization chromosomes from both parents come together. This means that the fertilized cell has the same number of chromosomes as the parent cell.

The fertilized cell or zygote divides by mitosis so that the same number and type of chromosomes are passed on to every cell. In this way the zygote grows and develops into a new plant or animal.

Key words

dominant gene	A gene which hides the effect of another gene in a hybrid
gene	Part òf the chromosome which controls inherited characteristics
hereditary characteristic	A characteristic inherited from the parents, e.g. eye colour
meiosis	A type of cell division. Pairs of chromosomes separate so that each sex cell has only half as many chromosomes as the parent cell
mitosis	A type of cell division. Each new cell formed by mitosis has the same number and type of chromosomes as the parent cell
recessive gene	A gene which does not appear to have any effect in a hybrid
zygote	A fertilised egg

Questions

1 Copy and complete these sentences:

   (a) The first person to explain inheritance by doing experiments with peas was a monk called ....

   (b) Inherited characteristics pass from one generation to the next in tiny thread-like structures called ....

   (c) A characteristic which disappears in the F_1 generation but reappears in about one quarter of the F_2 generation is called a .... characteristic.

   (d) A characteristic which appears in the F_1 generation and in about three quarters of the F_2 generation is called a .... characteristic.

2 Sex cells are formed when a cell divides by meiosis. What is the main difference between meiosis and mitosis?

3 Explain why when a mouse with a black coat was mated with a mouse with a white coat all their young grew black coats.

4 Copy and complete these diagrams to explain how eye colour is inherited. In each case say what colour eyes the children will have.

Parents

BB   Bb   Bb   bb

Gametes

At fertilization

5 How many chromosomes are there in human body cells?

6 Would a person with the chromosomes XY be a man or a woman?

7 A class did some breeding experiments with guinea pigs. A male guinea pig with a black coat was mated with a female with a brown coat. All their young had black coats. The same pair of guinea pigs were mated several times and their young always had black coats. Draw diagrams to help you explain these results.

20: Evolution

The book you are reading now looks very different from the first books. By looking in museums you can work out how the books used today have developed or **evolved**. These changes have taken place slowly and each type of book is slightly different from all the others.

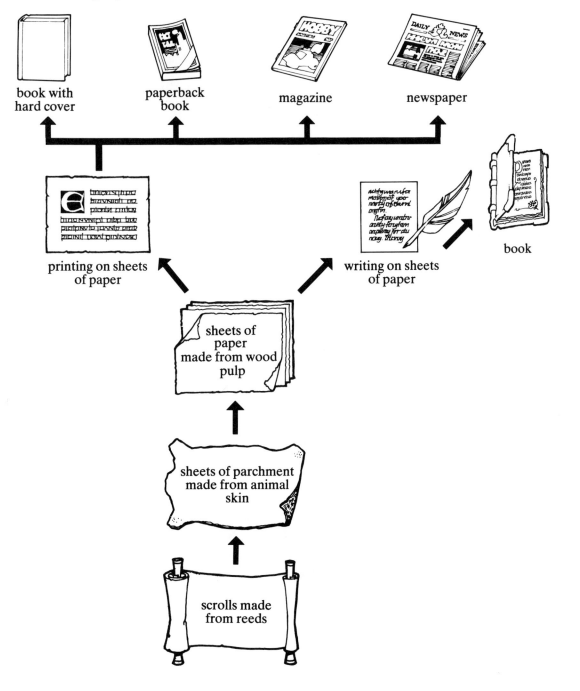

book with hard cover

paperback book

magazine

newspaper

printing on sheets of paper

writing on sheets of paper

book

sheets of paper made from wood pulp

sheets of parchment made from animal skin

scrolls made from reeds

Fossils

The surface of the earth is made up of layers of rock. The remains of some plants and animals have been preserved in the layers of rock. These remains are called **fossils**. By looking at fossils we can tell how plants and animals have evolved.

Some fossils were formed when insects became trapped in sticky resin from trees. Over millions of years the resin changed into **amber**. In this way, some insects have been preserved in amber. Other fossils were formed when plants decayed and left **a mould** or impression in coal.

The photograph shows a fossil formed when salts in the water turned the bones of a fish into stone.

Evolution of the horse

Fossils show how the horse has evolved from a small animal with four toes on its front foot. Over millions of years horses have become larger and developed longer legs with only one toe on its front foot.

today

Horse with only one toe on its front foot.

60 million years ago

Small animal with four toes on its front foot. Each toe had a hoof.

Evolution of the vertebrates

Fossils show that the vertebrates (animals with backbones) alive today evolved from fish which lived more than 400 million years ago.

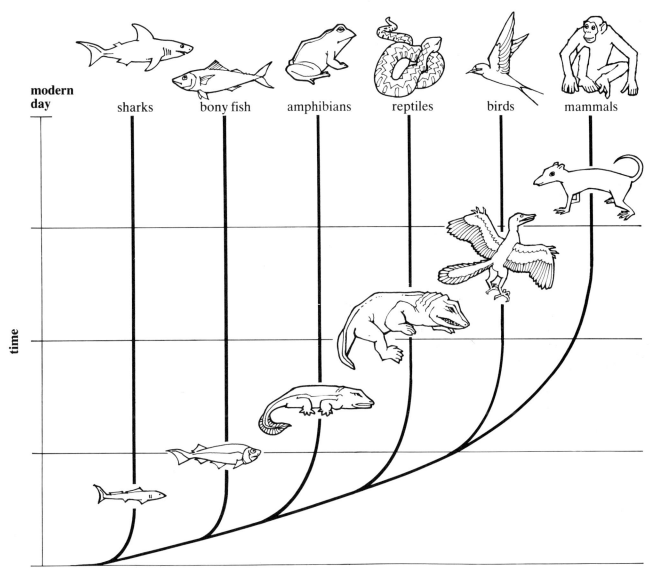

more than 400 million years ago

Questions

1 What are fossils?
2 Explain how fossils were made.
3 How does the horse today differ from the horse 60 million years ago?
4 The vertebrates alive today have evolved from the same type of animal. Which animal?

Gene mutation

Inherited characteristics are controlled by **genes** (see page 178). Sometimes a gene changes or **mutates**. When a gene mutates there may be a change in appearance or structure. Sometimes the change helps the plant or animal to live or **survive** longer than its parents. Sometimes the change makes it harder for the plant or animal to survive.

When a gene in a light coloured moth mutates the offspring are darker than the parents. In some places the dark colour helps a moth to survive. In other places the dark colour makes it harder for a moth to survive.

light coloured moth

gene for **light colour** passed on to offspring

gene **mutates**

Light coloured moths are easily seen in a smoky city. Many are eaten by birds. Only a few live long enough to reproduce.

Dark coloured moths are difficult to see in a smoky city. Only a few are eaten by birds. Many live long enough to reproduce.

only a **few light coloured** moths

many **dark coloured moths**

Questions

1 Explain what genes do.
2 What happens when a gene mutates?
3 Does a mutation always (a) help the offspring to survive or (b) make it harder for the offspring to survive?
4 Explain why there are more dark coloured moths than light coloured moths in smoky cities.

Natural selection

If there are light and dark coloured moths in a smoky city it is more likely that the dark moths will survive. This is known as **survival of the fittest** or **natural selection**.

Charles Darwin was the first person to suggest the idea of natural selection. He looked at the finches living on the Galapagos Islands. The finches on the islands were slightly different from those on the mainland of South America. Darwin suggested that all the finches on the islands had evolved from those on the mainland by natural selection.

Darwin said that if seed eating finches from the mainland were blown to the islands there would not be enough food for all the birds. Finches with slightly different beaks could eat different food. These finches would survive and produce young with the same type of beak.

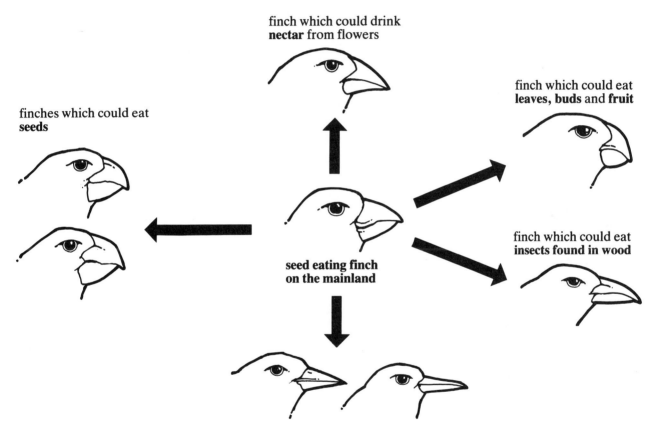

finch which could drink **nectar** from flowers

finch which could eat **leaves, buds** and **fruit**

finches which could eat **seeds**

finch which could eat **insects found in wood**

seed eating finch on the mainland

finches which could eat **small insects**

Artificial selection

Darwin's work explains how plants and animals evolve by natural selection. Man uses **artificial selection** to breed new varieties of plants and animals. Man has used artificial selection to produce many different plants, dogs, horses and pigeons from wild varieties.

Summary

By looking at fossils we can tell how life on earth has evolved. New types of plants and animals may eventually evolve from those alive today by gene mutation. If the mutation helps the offspring to survive longer than the parents the characteristic may be passed from one generation to another. Charles Darwin called this evolution by natural selection.

Key words

artificial selection   Breeding new plants and animals from other varieties

evolution   Slow change which results in a new type of plant or animal

fossil   Remains of a plant or animal which lived millions of years ago

gene mutation   When a gene changes to give the offspring a different characteristic to that of the parents

natural selection   Charles Darwin's idea that plants and animals which are suited to their surroundings will survive and reproduce

Questions

1 These fossils were found in different layers of rock. Each fossil is a leg bone of a horse. Put the fossils in their correct order, starting with the oldest.

fossil A   fossil B   fossil C   fossil D

2 Re-arrange these vertebrates to show the order in which they evolved.

man   **bony fish**   **reptile**   **amphibian**   **bird**

3 All the moths in the photographs belong to the same species. Birds eat both dark and light moths.

dark and light moths in a smoky area

 (a) Why do you think birds eat more:
   (i) dark moths in country areas
   (ii) light moths in smoky areas of cities.

 (b) Would you expect to find more light moths in a country area or in a smoky area of a city?

 (c) Explain how both light and dark moths could have evolved from the same type of moth.

 (d) Some cities have made smokeless zones (this means there is less smoke from chimneys). How do you think this has affected the number of:

 (i) dark moths (ii) light moths.

 Give reasons for your answer.

dark and light moths in a country area

4 What is the main difference between artificial selection and natural selection?

5 What is meant by natural selection?

6 Who first suggested that plants and animals evolve by natural selection?

7 Explain how the finches on the Galapagos Islands could have evolved from those on the mainland of South America.

8 What is meant by artificial selection?

9 How has man used artificial selection?

21: Soil and cycles in nature

Animals and plants depend on each other for food and other substances. Substances pass between living things and their surroundings or **environment**. Animals and plants are affected by the environment. The environment is affected by the activities of animals and plants. Soil is only one part of the environment which is affected by the activities of animals and plants.

Soil

Soil is being made all the time. Soil is made from the rock covering the earth's surface. The rock is cracked and broken down into smaller pieces by **weathering**.

Weathering is caused by:
1 changes in temperature e.g. strong sun and frost
2 the wind and rain

Small pieces of rock get worn down into smaller particles of gravel, sand, silt and clay. These are called **rock particles**. The rock particles get washed down into the river and sea beds.

How soil is made

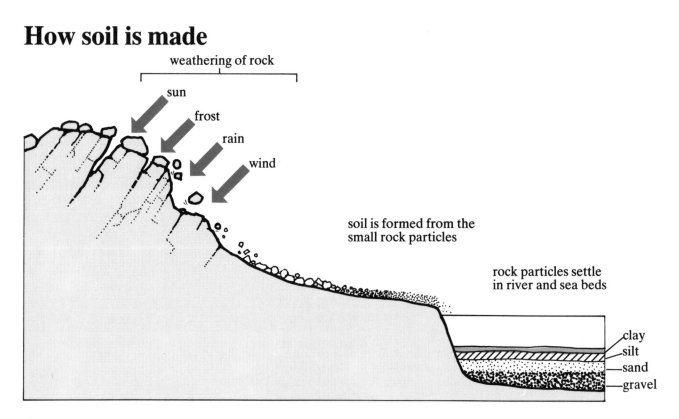

Soil structure

If soil is mixed and shaken with water the different parts of soil can be seen.

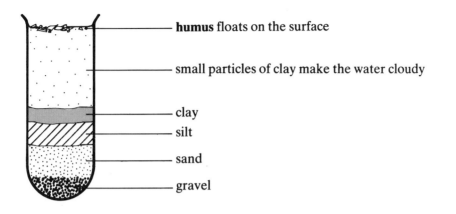

- **humus** floats on the surface
- small particles of clay make the water cloudy
- clay
- silt
- sand
- gravel

Humus

Humus comes from the waste substances and dead remains of plants and animals. The more humus there is in the soil the darker it looks. When bacteria feed on humus **nitrogen** and **other minerals** needed by plants are released into the soil. Humus helps to make **soil crumbs**.

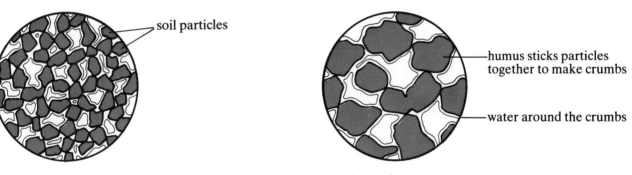

soil without humus

soil particles

small air spaces

soil with humus

humus sticks particles together to make crumbs

water around the crumbs

large air spaces

Questions

1 (a) How does the rock covering the earth's surface get cracked and worn away?
  (b) What is this process called?
2 What happens to the rock particles in river and sea beds?
3 How does humus help to make soil crumbs?

Types of soil

The structure of soil depends on the amount of clay, sand and humus in it.

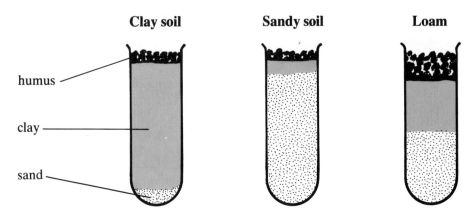

	clay soil	sandy soil	loam
soil structure			soil crumbs
size of particles	small	large	large
size of air spaces	small	large	large

Loam is the most **fertile** type of soil. The minerals needed by plants are not washed away when water drains through it. It is light and easy to dig.

Water drains quickly through a sandy soil. As the water drains through the soil the minerals needed by plants get washed away. If more humus is added the water takes longer to drain through the soil.

Water drains slowly through clay soil. In wet weather it is very sticky. In dry weather it forms very hard lumps. If more humus, sand and lime are added soil crumbs are made. This makes larger air spaces and helps the water to drain away.

Fertile soil

Plants grow well in fertile soils. A fertile soil is made up of:

1 rock particles
2 humus
3 minerals
4 water
5 air
6 living organisms (a) earthworms (b) fungi (c) bacteria

How earthworms improve the soil

Earthworms improve the soil when they make **tunnels** or burrows.

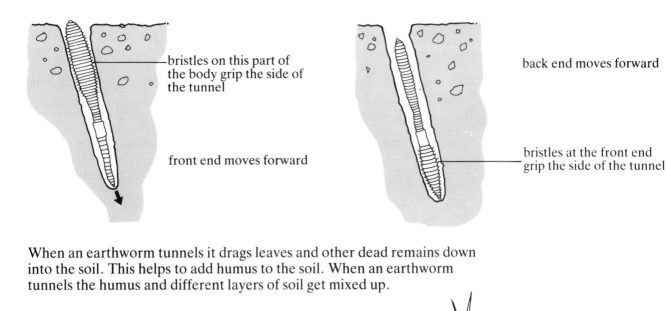

bristles on this part of the body grip the side of the tunnel

front end moves forward

back end moves forward

bristles at the front end grip the side of the tunnel

When an earthworm tunnels it drags leaves and other dead remains down into the soil. This helps to add humus to the soil. When an earthworm tunnels the humus and different layers of soil get mixed up.

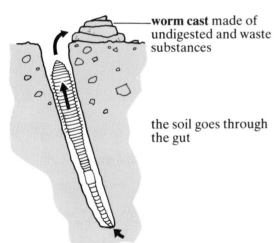

worm cast made of undigested and waste substances

the soil goes through the gut

the worm eats soil as it tunnels

water

air

plant roots

The tunnels help air, water and plant roots to go down into the soil.

Questions

1 Why are sandy soils light?
2 What happens when humus is added to a sandy soil?
3 Why is clay soil heavy and sticky?
4 How can clay soil be improved?
5 Which is the most fertile type of soil?
6 What is meant by a fertile soil?

7 Explain how an earthworm tunnels into the soil.
8 What happens to the soil when an earthworm tunnels?
9 How do earthworms help to let air and water into the soil?

How man improves the soil

Man can improve the fertility of the soil by:

1 **ploughing** the soil
2 growing **different crops** each year
3 adding **fertilizers**

Ploughing in the spring breaks up large lumps of soil. This helps the soil to dry out. It also lets more air into the soil.

Ploughing in the autumn mixes manure, and other things which will make humus, into the soil.

Crop rotation

Some plants take more minerals from the soil than other plants do. The nitrates taken from the soil by crops like wheat and turnips can be replaced by growing clover.

1st year	2nd year	3rd year

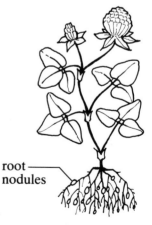

root
nodules

wheat takes a lot of minerals from the soil

enough minerals are left in the soil for **turnips**

clover has root nodules with bacteria in them. The bacteria change nitrogen into nitrates (see page 197).

The nitrogen cycle

Living organisms need nitrogen to make **proteins**. About 79% of the air is nitrogen but plants and animals cannot use this nitrogen. The nitrogen from the air must be changed into **nitrates** before it can be used by green plants. Animals get their nitrogen by eating plants.

The roots of some plants have swellings called **root nodules** on them. Bacteria living in the root nodules change nitrogen from the air into nitrates. Clover, pea and bean roots have root nodules.

This diagram shows what happens to nitrogen from the air and explains how nitrates are added to the soil.

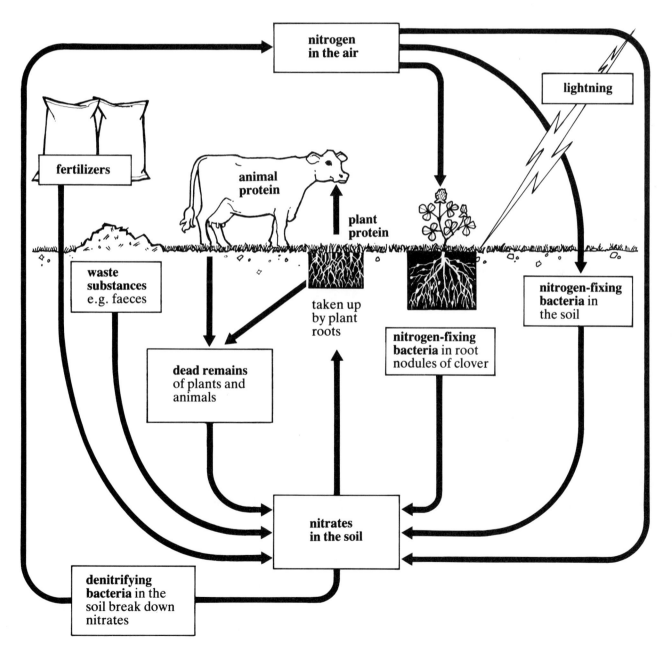

Questions

Look carefully at the diagram of the nitrogen cycle. Copy these paragraphs, filling in the spaces with these words:

1 nodules   bean   nitrogen-fixing   pea   soil

Nitrogen from the air is changed into nitrates by ………… bacteria. Some of these bacteria live in root ………… on clover, ………… and ………… plants. Others live in the …………

2 animals   proteins   roots

Nitrates are taken in by the ………… of green plants and used to make ………… Green plants are eaten by ………… In this way animals get nitrogen to make proteins.

3 fertilizers   denitrifying   waste   nitrogen   dead

Nitrates are added to the soil when ………… plants and animals and ………… substances such as faeces are broken down by bacteria. Sometimes farmers add ………… to the soil. ………… bacteria change nitrates into ………… gas.

The carbon cycle

Green plants get the carbon they need from carbon dioxide in the air. During the day green plants take in carbon dioxide and use it to make food. Carbon dioxide is returned to the air by respiration (see page 74) and decay.

This diagram shows how carbon passes from one living thing to another. It also explains why the amount of carbon dioxide in the air stays the same.

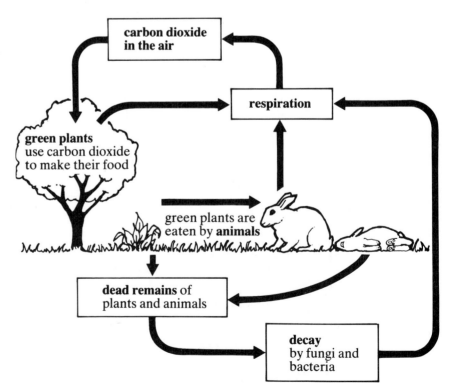

The water cycle

All plants and animals need water. The water from the sea **evaporates** and falls as rain on the land. Some of the rain water is taken in and used by plants and animals. Any water which is lost from plants and animals returns to the sea.

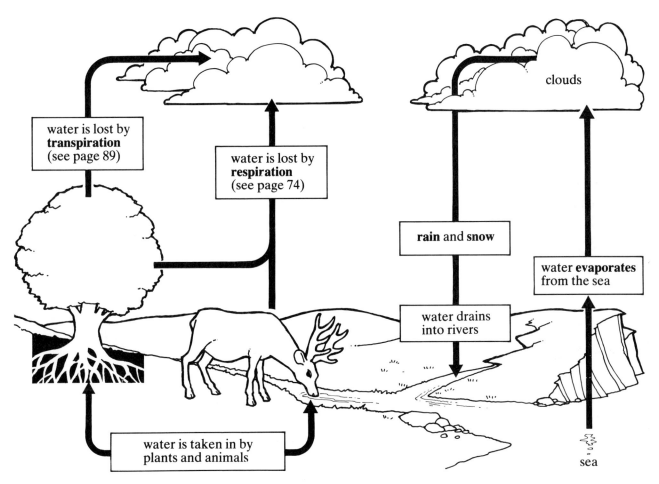

water is lost by **transpiration** (see page 89)

water is lost by **respiration** (see page 74)

clouds

rain and **snow**

water **evaporates** from the sea

water drains into rivers

water is taken in by plants and animals

sea

Questions

1 Why do plants need carbon dioxide?
2 How is carbon dioxide returned to the air?
3 Explain how the dead remains of plants and animals are broken down.
4 In your own words, explain how a rabbit gets the carbon it needs.
   (clue – start with carbon dioxide in the air)
5 What happens to the water evaporated from the sea?
6 Give two ways in which water is lost from plants.
7 How is water lost from animals?

Pollution

The nitrogen, carbon and water cycles show that substances pass between living things and the environment. If a substance which harms plants and animals gets into the soil, air or water it may be passed from one living thing to another. A substance which harms living things is called a **pollutant**. The chemicals sprayed on crops to kill insects and weeds are pollutants. The next diagram explains how these pollutants can pass from one living thing to another.

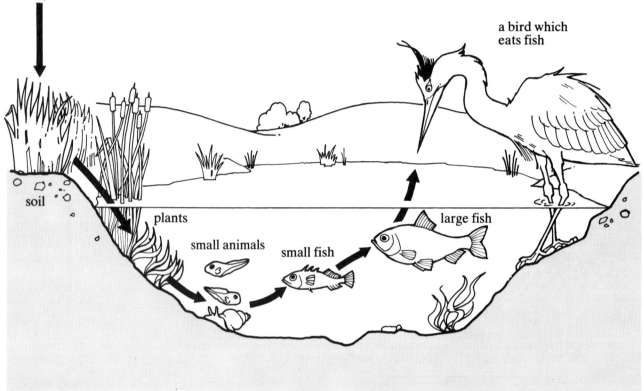

chemicals sprayed
on crops

a bird which
eats fish

soil

plants

small animals

small fish

large fish

Summary

There is a fine balance between the living and non-living parts of the environment. Living things affect the environment. The environment affects living things. Soil is one of the non-living parts of the environment. It is being made all the time from the rock covering the earth's surface. The structure and fertility of the soil can be improved by man.

Nitrogen, carbon and water are used time and time again as they pass from one living thing to another. If a pollutant gets into the soil, air or water it may be passed from one living thing to another.

Key words

denitrifying bacteria	Bacteria which break down nitrates and give out nitrogen gas
humus	Part of the soil formed from decaying remains of plants and animals
loam	Soil with a good mixture of sand, clay and humus
nitrates	Substances which contain nitrogen. They are taken in by plant roots and used to make proteins
nitrogen-fixing bacteria	Bacteria in the soil and root nodules which change nitrogen from the air into nitrates
pollutant	A substance which harms living things
weathering	How rocks are broken down into the smaller particles of soil e.g. by frost, sun, wind and rain

Questions

1 What is a pollutant?

2 Explain how a bird eating a fish could be harmed by a chemical sprayed on to a crop of wheat.

3 How does humus in the soil help plants to grow?

4 **(a)** How do plants with root nodules improve the soil?
    (b) Name a plant which has root nodules.

5 Why do some farmers grow different crops each year?

6 Explain why earthworms are good for the soil.

7 Name one living and two non-living parts of the soil.

8 Give three differences between clay and sandy soils.

9 Why do you think some farmers grow clover and then plough it into the soil before planting wheat in a field?

10 Which gas is given out by respiring organisms?

11 What percentage of the air is nitrogen?

12 Give two ways in which nitrates could be added to the soil (apart from using fertilizers).

13 Explain how nitrates are lost from the soil.

14 Carbon dioxide in the air is constantly being used by green plants. The amount of carbon dioxide in the air stays about the same. Explain how this happens.

15 In this experiment the same amounts of dry loamy soil, dry sandy soil and dry clay soil were used. The same amount of water was added to each type of soil.

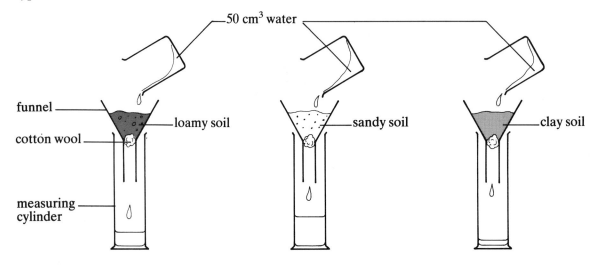

This table shows the results of the experiment.

	loamy soil	sandy soil	clay soil
time taken for first drop of water to fall into the measuring cylinder	70 seconds	12 seconds	305 seconds
volume of water which passed through in 30 minutes	25 cm^3	40 cm^3	6 cm^3

(a) Explain why the time taken for the first drop of water to fall into the measuring cylinder was different for each type of soil.

(b) Why was the volume of water which passed through each type of soil different?

(c) Why do you think cotton wool was put into each funnel?

(d) What can gardeners with clay soil do to help drain their soil?

(e) What would each of the soils be like to dig:

  (i) after heavy rain (ii) after a long dry spell in the summer?

(f) What could a gardener add to sandy soil to stop the water draining through too quickly?

16 In this experiment bicarbonate indicator has been used to make five 'miniature ponds'.

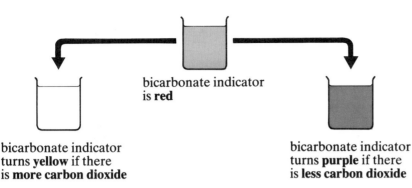

bicarbonate indicator
is **red**

bicarbonate indicator
turns **yellow** if there
is **more carbon dioxide**

bicarbonate indicator
turns **purple** if there
is **less carbon dioxide**

These five tubes were left in sunlight for one hour. The bicarbonate indicator in each tube was red at the start of the experiment.

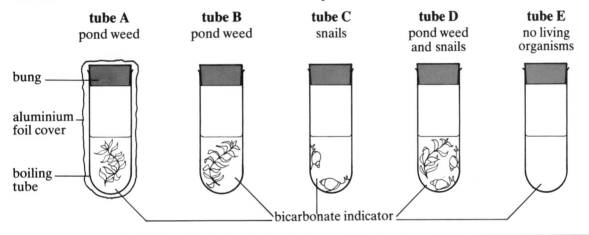

Results after one hour					
	tube A	**tube B**	**tube C**	**tube D**	**tube E**
organism	pond weed	pond weed	snails	pond weed and snails	none
conditions	dark	light	light	light	light
colour of indicator at start	red	red	red	red	red
colour of indicator after one hour	yellow	purple	yellow	red	red

(a) Why did the bicarbonate indicator in tube E stay red?
(b) Why did the bicarbonate indicator in tube C turn yellow?
(c) Why did the bicarbonate indicator in tube B turn purple?
(d) Why did the bicarbonate indicator in tube D stay red?
(e) If more snails were added to tube D would the bicarbonate indicator turn purple or yellow or stay red? Explain your answer.
(f) Why did the bicarbonate indicator in tube A turn yellow?
(g) Why was tube A covered with aluminium foil?

22: The variety of life

There are three main groups of living things:

1 simple organisms
2 plants
3 animals

Simple organisms

Protozoa

Protozoa are **unicellular** organisms (have only one cell).

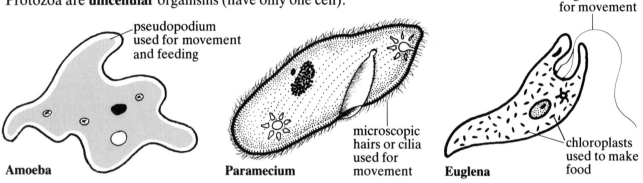

Amoeba — pseudopodium used for movement and feeding

Paramecium — microscopic hairs or cilia used for movement

Euglena — whip-like flagellum used for movement; chloroplasts used to make food

Fungi

Fungi **cannot make their own** food because they do not have chlorophyll (green colour). Some fungi are **parasites**. Parasitic fungi feed on other living organisms. Ringworm in animals and mildew in plants are caused by parasitic fungi.

Some fungi are **saprophytes**. Saprophytic fungi feed on dead and decaying material.

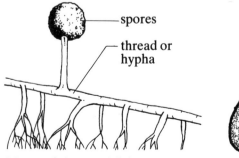

Mucor (pin mould) is a saprophytic fungus which feeds on bread.

spores

thread or hypha

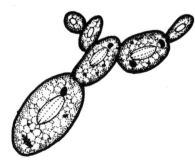

Yeast cells

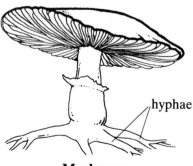

Mushroom

hyphae

Algae

Algae can **make their own food.**

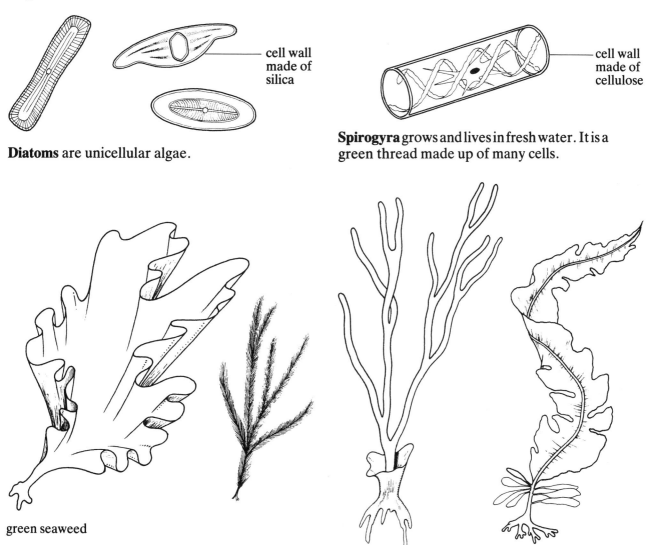

cell wall
made of
silica

Diatoms are unicellular algae.

Spirogyra grows and lives in fresh water. It is a green thread made up of many cells.

cell wall
made of
cellulose

green seaweed

Seaweeds are algae which grow in the sea.

brown seaweed   red seaweed

Questions

1 (a) Name a unicellular organism which has a pseudopodium.
   (b) What is a pseudopodium used for?
2 Name a plant disease caused by a fungus.
3 Which type of organism causes ringworm in animals?
4 What do saprophytic fungi feed on?
5 What are diatoms?
6 Would you expect to find Spirogyra living in a pond or in the sea? Give reasons for your answer.

Plants

There are two main groups of plants:

1 plants which do not have flowers (**non flowering plants**)
2 plants which have flowers (**flowering plants**)

Non-flowering plants

Mosses and liverworts

Mosses and liverworts do not have flowers. They reproduce by **spores**. They live in damp, shady places.

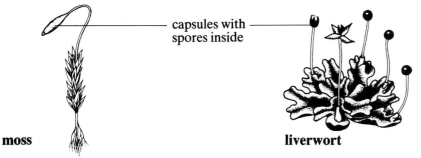

capsules with spores inside

moss

liverwort

Ferns and horsetails

Ferns and horsetails do not have flowers. They reproduce by **spores**.

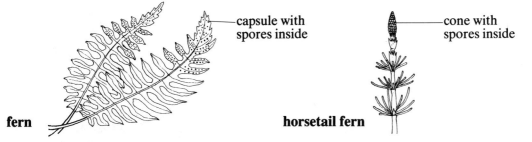

capsule with spores inside

cone with spores inside

fern

horsetail fern

Plants which have cones

Trees which have cones are called **conifers**. They reproduce by **seeds** which are inside the cones.

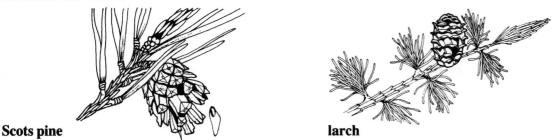

Scots pine

larch

Flowering plants

The **reproductive organs are in the flowers**. The male sex cells or pollen grains are made in the anthers. The female sex cells or ova are made in the ovary. After fertilization a fertilized ovum develops into a **seed**. The ovary wall develops into a **fruit**. (see page 152).
A plant which grows from a seed with only one seed leaf or **cotyledon** is called a **monocotyledon**. A plant which grows from a seed with two seed leaves is called a **dicotyledon**.

Monocotyledons

These plants have leaves with parallel veins and long thin leaves.

meadow grass   **maize**   parallel veins

Dicotyledons

This is the largest group of flowering plants. Their leaves have a network of veins.

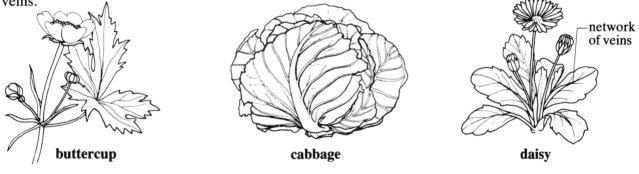

buttercup   **cabbage**   **daisy**   network of veins

Questions

1 What are the two main groups of plants called?
2 Which plants reproduce by spores?
3 Which plants reproduce by seeds?
4 What does a fertilized ovum develop into in a flowering plant?
5 What does the ovary wall develop into in a flowering plant?
6 Give two differences between monocotyledons and dicotyledons.

Animals

Animals can be divided into two main groups:

1 animals without backbones are called **invertebrates**
2 animals with backbones are called **vertebrates**

Animals without backbones

Coelenterates

Coelenterates have a **hollow body**. The mouth is the only opening to the body. The mouth is surrounded by **tentacles**. There are **sting cells** on the tentacles. The sting cells paralyse small animals.

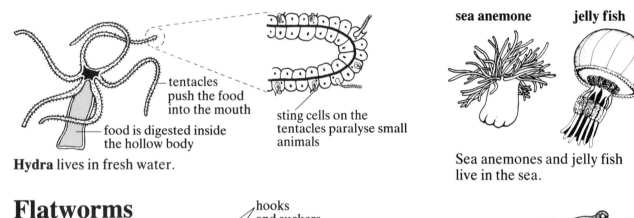

tentacles push the food into the mouth

food is digested inside the hollow body

sting cells on the tentacles paralyse small animals

Hydra lives in fresh water.

sea anemone   jelly fish

Sea anemones and jelly fish live in the sea.

Flatworms

hooks and suckers

sucker

Flatworm which lives in fresh water.

Tapeworm which lives inside the gut of man.

Liver fluke which lives inside a sheep's liver.

True worms

True worms have bodies made up of rings or **segments**.

sucker

An **earthworm** lives in the soil.

A **leech** can suck blood.

A **lugworm** lives in the sea.

Arthropods

Arthropods have a hard layer on the outside of the body. This is called an
exoskeleton (see page 108).
The arthropods are divided into smaller groups.

Crustaceans

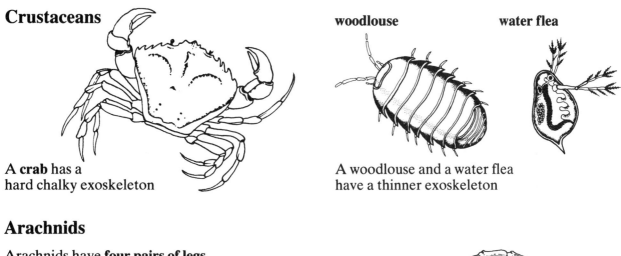

woodlouse   **water flea**

A **crab** has a
hard chalky exoskeleton

A woodlouse and a water flea
have a thinner exoskeleton

Arachnids

Arachnids have **four pairs of legs**.

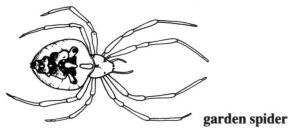

garden spider

scorpion

Insects

Insects have **three pairs of legs**. The **body** is in **three parts**.

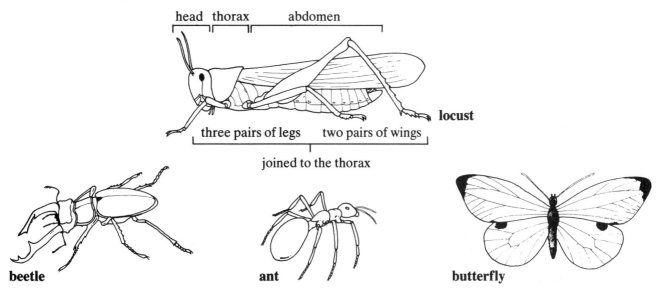

head   thorax   abdomen

locust

three pairs of legs   two pairs of wings

joined to the thorax

beetle   **ant**   **butterfly**

Centipedes and millipedes

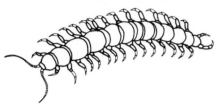

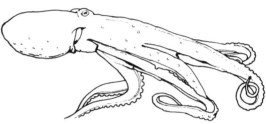

Centipedes have one pair of legs on each segment.

Millipedes have two pairs of legs on each segment.

Molluscs

Molluscs have a **soft body** and one or two **shells**.

A **snail** has one shell.

A **mussel** has two shells.

An octopus has a shell inside its body.

Spiny-skinned animals

Spiny-skinned animals live in the sea. They have a thick **spiny-skin**. The body is in **five parts**.

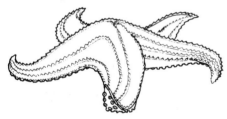

A **starfish** has suckers which are used for moving from one place to another.

In a **sea urchin** the five parts of the body are joined together.

Questions

1 Give another name for animals without backbones.
2 What are sting cells in coelenterates used for?
3 Name two flatworms which are parasites (live inside other living organisms).
4 Which group of invertebrates have bodies made up of rings or segments?
5 Name a group of animals which has an exoskeleton?
6 How many legs has a spider?
7 How many legs has an insect?
8 Name the three parts of an insect's body.
9 What is the main difference between a centipede and a millipede?

210

Animals with backbones

Animals with backbones are called **vertebrates**. There are five groups of vertebrates.

1 fish   **3** reptiles   **5** mammals
2 amphibians   **4** birds

Fish

Fish **live in water**. They have a **streamlined body** which is covered with **scales**. Their fins are used for movement and balance. Most **lay eggs**. They breathe with **gills**.

cod

herring

A **shark** has a skeleton made of cartilage.

The cod and herring have a skeleton made of bone.

Amphibians

Amphibians have **moist skins.** They **lay** their **eggs in water**. The eggs hatch into **larvae** which breathe with **gills**. The larvae change into **adults** which breathe with **lungs**. Adult amphibians can live on land.

frog   **toad**   **newt**

Reptiles

Reptiles have **dry scaly skins**. Most **lay eggs** with tough leathery shells. They breathe with **lungs**.

 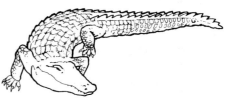

A **tortoise** and **adder** live on the land.

Crocodiles live in water but can come onto land.

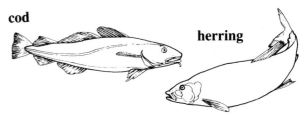

211

Birds

Birds have **feathers** on their bodies. They **lay eggs** with hard shells. All birds have **wings**.

eagle

duck

The eagle and duck use their wings for flying.

The **penguin** has wings but cannot fly.

Mammals

Mammals have **hair or fur** on their bodies. Female mammals have **mammary glands** (or breasts) which make milk. **Young mammals suck milk from their mothers** until they are old enough to feed themselves. Mammals can be divided into three groups:

1 mammals which lay eggs
2 mammals which have pouches
3 mammals which have a placenta and give birth to live young

Egg laying mammals

These are primitive mammals. They **lay eggs** in a burrow. When the young hatch they feed on their mother's milk.

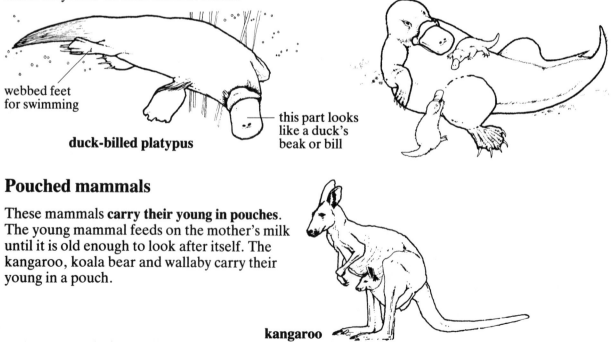

webbed feet
for swimming

duck-billed platypus

this part looks
like a duck's
beak or bill

Pouched mammals

These mammals **carry their young in pouches**. The young mammal feeds on the mother's milk until it is old enough to look after itself. The kangaroo, koala bear and wallaby carry their young in a pouch.

kangaroo

Placental mammals

These mammals develop inside the mother's body. Before the young mammal is born it is joined to the mother's womb by the umbilical cord and **placenta**. The young mammal gets food and oxygen through the placenta.

After birth the young mammal feeds on its mother's milk.

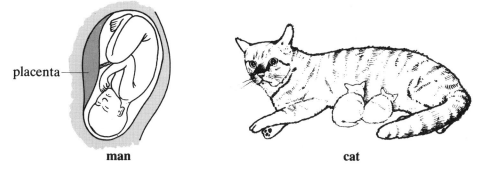

placenta

man

cat

The body temperature of animals

Invertebrates, fish, reptiles and amphibians are called **cold blooded animals**. Their body temperature changes as the temperature of the surroundings changes. Cold blooded animals may die if their surroundings are too hot or too cold.

Birds and mammals are called **warm blooded animals**. In warm blooded animals the body temperature stays about the same even in very hot or very cold surroundings.

Questions

1 Name the five groups of vertebrates.

2 Which animals breathe with gills?

3 Give one difference between the skin of an amphibian and the skin of a reptile.

4 Give three characteristics of mammals.

5 **(a)** Which is the most primitive group of mammals?
   (b) Give the name of one mammal in this group.

6 Name two mammals which carry their young in pouches.

7 **(a)** Which group of mammals develop inside the mother's womb?
   (b) How do they get their food and oxygen?

8 What are young mammals fed on after birth?

9 **(a)** What is meant by a cold blooded animal?
   (b) Name four groups of cold blooded animals.

10 **(a)** What is meant by a warm blooded animal?
   (b) Name two groups of warm blooded animals.

Summary

Plant groups

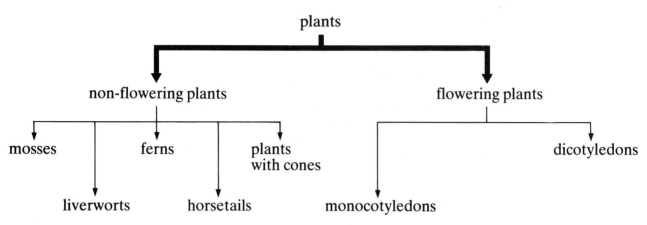

Animal groups

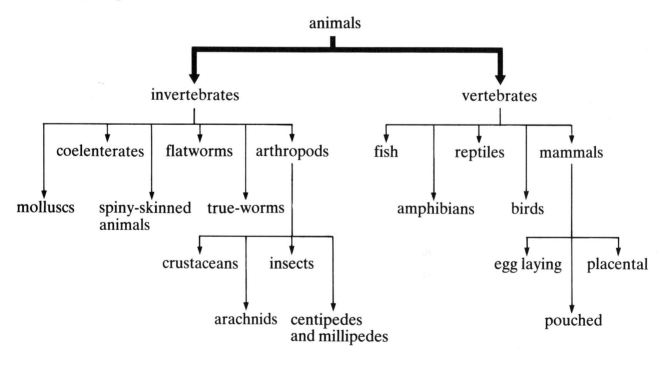

Key words

cold blooded animal   The body temperature changes as the temperature of the surroundings changes

invertebrate   An animal which does not have a backbone

unicellular   Only one cell

vertebrate   An animal which has a backbone, e.g. fish, amphibian, reptile, bird, mammal

warm blooded animal   The body temperature stays the same even if the temperature of the surroundings changes, e.g. bird and mammal

---

Questions

1 Choose one word from each list to complete these sentences.

(a) An example of a saprophytic fungus is............

moss, seaweed, Amoeba, pin mould, fern.

(b) An example of a parasitic fungus is............

Spirogyra, robin, mildew, Euglena, liverwort.

(c) Plants which have a network of veins in their leaves are called............

cotyledons, monocotyledons, dicotyledons.

(d) One example of a cold blooded animal is the............

penguin, crocodile, whale, polar bear.

(e) One plant which has seeds is............

moss, fungi, fern, grass.

(f) Animals with backbones are called............

vertebrates, crustaceans, vertebrae, invertebrates.

(g) Scales cover the body surface of all............

amphibians, birds, vertebrates, fish, mammals.

2 Give two reasons for saying that the rabbit is a mammal.

3 Name a mammal which lives in water.

4 Give two differences between reptiles and amphibians.

5 Which type of organism causes ringworm in man?

6 Give two differences between a mushroom and a buttercup.

7 Explain what is wrong with saying "only fish breathe with gills"?

8 Copy and complete the following table to show the differences between the groups of vertebrates.

	mammals e.g. man	birds e.g. robin	reptiles e.g. tortoise	amphibians e.g. frog	fish e.g. herring
What type of skin do they have?	hairy				
Are they cold blooded or warm blooded?					
Do they breathe with lungs or gills?		lungs			
Do they lay eggs?					yes
Do they have fins, wings or legs for moving?					

9 Copy the summary diagrams showing
   (a) the groups of non-flowering plants
   (b) the invertebrates
   (c) the vertebrates.

Draw a picture of one organism in each group you have mentioned.

Index